what i wish my roofer had told me

The Ultimate Guide to the Roof of Your Dreams on a Budget

jon nelsen

life level up books, llc

Copyright © 2022 by Jon Nelsen

Copyright © 2022 by Life Level Up Books, LLC

All rights reserved.

Disclaimer Notice:

Please note the information contained within this document is for educational and entertainment purposes only. All effort has been executed to present accurate, up to date, reliable, complete information. No warranties of any kind are declared or implied. Readers acknowledge that the author is not engaged in the rendering of legal, financial, medical or professional advice. The content within this book has been derived from various sources. Please consult a licensed professional before attempting any techniques outlined in this book.

By reading this document, the reader agrees that under no circumstances is the author responsible for any losses, direct or indirect, that are incurred as a result of the use of the information contained within this document, including, but not limited to, errors, omissions, or inaccuracies.

This book is written for entertainment purposes only. The statements made in this book do not necessarily reflect the present market at the time of reading or current views of the author. Furthermore, the author accepts no responsibility for actions taken by the reader as a result of information presented in this book.

No part of this book may be reproduced in any form or by any electronic or mechanical means, including information storage and retrieval systems, without written permission from the author, except for the use of brief quotations in a book review.

contents

Introduction	ix
The Roof *Like Living in a Cave Only More Expensive*	1
8 Reasons You Need to Fix Your Leaky Roof Now	3
How Long Does a Typical Roof Install Take and Tips to Avoid the Worst Mistakes *What You Need to Do Prior to the Installation of New Roofing*	9
The Best Energy Efficient Roofing *What You Need to Know About the Different Types of Roofing Materials and Roof Products*	15
Will a New Roof Reduce Your Insurance Premiums? *How a New Roof Can Lower Insurance Premiums and Other Tips to Lower Homeowners Insurance*	25
Spring Cleaning Roof Checklist	31
How Do I Find a Good Roofing Contractor?	39
17 Questions to Ask a Potential Roofing Contractor *What Do You Need to Look For When Hiring a Local Roofer*	45
How Often Should You Clean Your Gutters? *The Benefits, Why It Matters, and if You Might Need a Professional Gutter Cleaner*	52
Final Words	60
Bonus: Solar Powered Energy Theft Preview *Chapters 1-3*	62
Better Energy *Where Energy Comes From & Why Residential Solar Just Makes Sense*	64
Panels + System *How Solar Panels & the System Are Designed and Function*	83
Is My House Right for Solar? *Find Out if Your House Makes the Cut*	90
Also by Jon Nelsen	101

WHAT I WISH MY ROOFER HAD TOLD ME

The Ultimate Guide to the Roof of Your Dreams on a Budget

JON NELSEN

SOLARTURNSMEON.COM

STOP RENTING YOUR ENERGY

Switch to solar for $0 down and save up to $200/mo on electric bills. Increase your home value and get a 30-year warranty.

introduction

Of all our basic needs, shelter seems to be the one that is forgotten about the most. Sure, we all have a home to live in, but is it the best it could be? Is our home what we dream of when we close our eyes and does it bring a smile to our face when we pull into the driveway?

For curb appeal, dollar for dollar, there are few upgrades you can make with more impact than a solid roof. When we look at a home with a well-crafted roof, it inspires us to imagine the inside is just as immaculate and breathtaking. The opposite is, of course, true as well. A discolored, ugly roof begs the assumption that we might not want to see what's behind those closed doors.

Hiring a contractor to replace or repair your existing roof has, for decades, remained purposefully mysterious. This book hopes to illuminate some of the mystery of a hiring a contractor to replace that worn down roof of yours and help you avoid the pitfalls that can cost you and your family thousands.

Introduction

Even at a young age, I was helping my father replace roofs on aging homes, restoring them to the beauty they once had. I hope to provide a simple resource that allows you, the homeowner, to make an informed decision about the most important and potentially expensive aspect of your home!

This book tries to answer most of the common questions relating to both roofing and solar. While this may not be the most exciting book you read this year, the information might just be the most important!

Introduction

the roof

. . .

Like Living in a Cave Only More Expensive

IF YOU'RE READING this book, chances are high you are just starting your journey into the magical land of roof repair and replacement. Except for paying a dentist to pull a tooth, I can think of no more annoying way to spend hard earned money. Unfortunately, like that tooth example, the longer you wait to do the job, the worse it will hurt in the end.

YOUR ROOF IS ALL that stands between you and the elements and so it's in your best interest to spend generously to ensure you have a quality product hanging above your head. But simply cutting a check isn't enough. If you really want to get your money's worth from that roof, you need to maintain it. Year after year and season after season, you need to focus on preventative maintenance to ensure your roof's longevity. While I can't save you from the cash you will spend on your roof, I can try to help that cash go farther.

THIS BOOK WILL DISCUSS how to hire the best roofer in your area, decide the best material for your roof and discuss the preventative maintenance you need to do and when you need to do it. It's filled with actionable tips to save you money and help eliminate present or future headaches.

WHILE THIS BOOK is a short sweet easy-to-read manual to all things roofing, if you know what you need, skip to that chapter and use this as a reference guide. Hopefully, this book saves you thousands on your next roofing job and prepares you to avoid some pitfalls that often come with a large construction project. Because as the great Benjamin Franklin said:

"By failing to prepare, you are preparing to fail."

8 reasons you need to fix your leaky roof now
. . .

SO LET'S get this out of the way right now, if you have a roof leak you should be on the phone with a roofer and not reading this book. But, if you're still here, I would like to share how to recognize leaky roof issues so that you can fix them before they turn into a bigger more expensive problem.

What do you need to know about roof leaks?

1. **Common causes of leaks**
2. **Signs of leaks**
3. **Costs to fix a leaky roof**
4. **What that weird smell means**

A LEAKY ROOF is more than just a nuisance! Without prompt attention, a leaky roof can quickly turn into an expensive headache, costing you time, money, and most times the health of you and your family. While the problem may be small now, in a few weeks or even days, it can turn into a dangerous situation. The weather around North America is as varied as the people. Here in the northeast we all love the changing of the seasons. However, your roof doesn't.

The Danger of a Leaky Roof

WHEN A ROOF LEAK causes moisture to spread inside your home, it can lead to serious health problems for your family. From mildew and mold to serious health problems such as dehydration, respiratory issues, and even cancer, having an unhealthy roof can have serious implications.

Why is Your Roof Leaking?

DIFFERENT FACTORS CAUSE LEAKY ROOFS, and most times it's a combination of things that take a while to occur. Sometimes a leaky roof results from damage from rain and storms, perhaps even ice or a burst pipe. Other times, the source is from problems with the gutters and downspouts.

THERE ARE a few different causes of roof leaks and each one needs to be addressed, so the leak doesn't continue or become worse. Perhaps the cause is as simple as failing to change the roof drain when your gutters were cleaned last. If that's the case, let a pro worry about climbing those tall ladders and fixing the gutters. If the leak isn't a result of water coming from the gutters but has something to do with your roof, there may be a leak in your roof foundation and the structure will need to be reinforced. In some cases, the roof may have become cracked due to age or weather abuse and needs to be completely replaced. If a leaky pipe caused the leak, the building owner might even be required by law to conduct a roof inspection to determine if the damage is severe enough to require replacement of the entire roof.

What the cost of a roof leak really is

A WATER LEAK is anything that has water dripping down from the roof to the interior structure. This can be from leaks inside the building from the roof gutter or it can be a leak coming through the attic. An average leak will cost you anywhere from $100 to $2,500. It's also important to note that the bigger the leak, the more you'll likely need to pay. A leak that comes through a hole in the roof might only cost you $400 or less. However, if it's from the roof lining, and leaks through 2nd or 3rd level walls, that estimate can jump to thousands! The cost of a leak can also quickly escalate to more than just dollars and cents. Most roof leak problems stem from the cracks and cavities that exist in your roof, which left untreated, can lead to invisible mold spores in your home that can cause respiratory issues for people who live there.

Signs that you need to fix your roof

- You might notice the ceiling dripping down after a heavy rainfall. Or the water from the gutters will seep onto your property, ruining flower beds or ponding in your driveway rather than diverting away from the home.
- Your family and pets can't play outdoors because the water can pool up on the ground without moving, which can show the beginning of a foundation issue.
- The ground around your home may have an unpleasant odor due to mold or mildew.

- The roof will rust or discolor and most times, the integrity of it will deteriorate. If you spot these things early, you can avoid further damage.

THE TIME TO act before it is too late. A leaky roof is more than just a nuisance! Without prompt attention, a leaky roof can spin into a costly headache.

A wet carpet or ceiling

HUGE OR MINOR lacerations on your ceiling may be due to:

- Heavy rainfall
- Extended hot summer days, or frosty nights
- Rains and lighting storms
- A faulty or failing HVAC system
- Freezing and thawing

and more...

IF THESE SCENARIOS sound like the cause of your roof leaks, it might be time to have a professional inspection before it causes further damage to your home. The first step is to perform a careful visual inspection of your roof to see if there is anything you can do about the problem yourself, or if you need to call a professional to come out. Sometimes, a good hard rain can trigger a leak, but the best solution is to catch it while it's dry so you can get it fixed quickly and efficiently.

Mold on the walls or in the attic

Are you concerned about a musty smell inside the attic or upper rooms of your house or your house smells as if it hasn't aired out properly? However, you've looked up there and did not see anyone there. It's probably mold. Fortunately, there are simple things you can do to make sure it stays trouble free. Check for minor cracks and leaks. Even a small crack in the ceiling can get larger quickly. Look carefully and make sure that nothing seems amiss. If you are finding cracks on the siding, especially under porches, make sure you get them repaired before the minor problem becomes a large one.

A musty smell in the house

Sometimes a leak in the attic may find its way to lower foundations of the house via pipes and interior walls. A crack in the kitchen's wall might lead to mildew in the walls and lifting of your new tiles. Ceiling and floor tiles damaged by water show leaks, but often once you notice it, that indicates it's already become a larger issue.

Perhaps it's noisy creaks and vibrations from the roof you chalked up to an aging home, making it impossible to sleep. A separation between joints causes noises and creaks within the house, allowing moisture to penetrate. Which, of course, can lead to mold and mildew, making your entire house smell like a wet dog. Let us send a roofing specialist to investigate any potential hazards that may be present.

How to mitigate water damage from a leaky roof

BEFORE YOU DECIDE to add this to a honey do list, take a moment to consider the worst-case scenario. In a worst-case scenario, what would happen if your small leak turns into a major problem? A problem affecting the health of your family or the foundation of your home? If you haven't taken the proper steps to protect your roof in the past, you're going to want to get started now. Let us send out a professional to look for places where water can come in. This means checking the front and back decks, chimneys, windows, and gutters. Since your roof is one of the most important things protecting you and your family from the elements, it's critical that you keep water from getting in. We can investigate areas that could leak from your gutters or downspouts, or even from your attic, and treat those areas with a preventative maintenance sealant to keep water out.

Fixing leaks as soon as they happen

IF YOU HAVE a leak in your roof, it is best to fix it right away. Not only do small leaks turn into larger ones, they can also become more difficult to fix if you wait too long. A leaky roof can be subtle, but it can still make a mess in your home and on your property. A few drops of water is all it takes to start a small rainstorm inside of your home. Although water may seem insignificant, water droplets that fall on a flat roof become very heavy and can eventually cause your roof to collapse or crumble.

how long does a typical roof install take and tips to avoid the worst mistakes

. . .

What You Need to Do Prior to the Installation of New Roofing

SO WANT to know how long a typical roof install takes? Short answer, it typically takes around 1-3 days to install a roof, depending on the size of the roof and what's needed. Not so bad huh, many homeowners worry that installing that new roof will take longer than it really does. That's because homeowners are often used too long and costly home remodels or repairs that run over budget. Keep in mind that installing skylights on the roof or installation of a new gutter system can lengthen the time your contractor will need to complete the work. Likewise, how to install a skylight on a metal roof differs from how to install a skylight on a shingled roof. The good news is that while a new roof is an enormous project, it is also a relatively simple one that your builder has done many times before. Straightforward if you are dealing with an experienced roofing professional who has dealt with all obstacles to the completion of the job. But before jumping out of your chair and dialing the nearest roofer and committing to a major home repair or replacement, other important factors must be considered. Let's discuss the best way to prepare for a new roof and prepare your home for the crew that will install it.

Prepare Your Family - Think About Pets and Children

WHILE STAYING out of potential dangerous zones of your home during the construction or repair of a new roof might seem like common sense to you, some members of your home might not even realize such dangers exist. Take the time to explain to everyone in your house that certain areas of the house will be a "no go" zone while the crews are doing work in your home. Areas to avoid include the upper levels of the house and certain areas of the yard and driveway. It's critical to steer clear as they will fill the yard with dangerous debris such as nails, broken pieces of wood, and old shingles, all of which can cause serious injury to a pet's paws or small children. In addition, one accidental bump into a ladder could mean a trip to the hospital for the crew working above. It's often best to move most of the family to friends, family or neighbors as much as possible during the work.

YOU WILL PROBABLY WANT to keep one member of the house present in case there are questions the roofers have, so someone can answer immediately to keep the work flowing. Just be prepared for loud noises and a strong possibility of disturbances to normal sleeping routines. This is because roofers typically like to begin early and stay late to get the job done as quickly as possible.

Moving Your Vehicles

OFFERING the work crews the closest most accessible location at the home gives you several benefits and makes for efficient and happy contractors. First, you don't want your vehicle to get hit by the falling debris that accompanies a roofing job. Second, the crew will need constant access to vehicles for tools, supplies, water and food breaks, along with the need to run to the nearest roofing section of Home Depot in case something they need didn't make it to the job site. It's also a good idea to keep the doors of your garage or shed closed to avoid having additional cleanup of dust and debris.

Use Tarps and Blankets to Cover Belongings in the Attic

WHILE MOST OF the junk we end up storing in the attic isn't on our priority list, keeping cherished family heirlooms or photo books safe should be. The constant pounding on the roof and workers walking around will undoubtedly knock down dust and dirt onto whatever is directly beneath it. In some cases you might have upper rooms of the house you wish to secure from dirt as well. Old blankets, sheets, or drop clothes make efficient covers of valuables during a roofing job.

To the Windows to the Walls

No, it's not a song from your college days. It's a reminder to secure valuables that are at risk of crashing because of the vibrations that will happen in your home. Paying careful attention to the upper floors of your home, carefully go around and remove anything on the walls that isn't screwed in. Think about tall shelves that might have knick-knacks that are at risk of a tumble from wall vibrations. Framed pictures should be taken down and you might even consider removing chandeliers and other decorative light fixtures.

For the Love of God, Move the Grill!

BEFORE THE ROOFING work even begins, you will want to ensure that deck and patio furniture have been placed in a garage or shed. If you're lacking in storage or your garage is too full of junk (looking at you, grandpa John), move your items to the furthest point in the yard away from the work. And unless you want your expensive Big Green Egg to go Humpty Dumpty, make sure grills and grill equipment are moved away from the base of the house. There is nothing worse than having to replace a perfectly good grill after shelling out your hard earned money for a new roof. Then again, a beautiful new roof and shiny new grill has a nice ring to it, don't you think?

Mow the Lawn or Be Prepared to Step on Old Shingles and Roofing Nails

CUTTING the grass shorter than usual will help roofing contractors identify debris that may have fallen during the cleanup stage. Even though most roofers will use drop clothes, it's a good idea to make their job as easy as possible. The last thing you want is to step on a rusty roofing nail that was left in your yard under tall grass. It also is a good idea to trim large trees or bushes that are near the base of the house that you may have been putting off. Finally, it should go without saying if there are trees or branches in the roof's vicinity, cut those back as well to make the roofing area clear and visible.

Does Your Roof Have Old Antennas or Satellite Dishes?

If you have old unused satellite dishes or antennas on your roof, let the contractor know so they can remove them. Most will be happy to take them away with the rest of the old roofing they are disposing of. If you have satellite dishes or antennas in use, let them know that too. It is also a good idea to contact your cable or satellite company and ask the best way to handle a roof repair. They may have suggestions or be willing to send someone out to help things go smoothly. We have even heard of customers getting upgraded satellite dishes from cable companies free of charge. It never hurts to ask.

It is also a good idea to talk to your roofer and see if they will reuse old roof vents or if there is additional cost to install a new roof vent. The cost to install roof vents may not be that great, but it may extend the life of your new roofing and help lower cooling costs during the height of the summer heat.

Identify the Best Outlets for Your Roofer

Chances are that whatever roofing company you hire will want to know if there are any exterior outlets they can use during the roof replacement or repair. If you do not have any easily accessible outlets outdoors, perhaps one from the garage could work? Otherwise you have to crack a window and let the contractor run extension cords from inside the house, which of course can sometimes be a hassle for everyone involved.

Remember to Tell Your Neighbors

While it might not seem like a big deal, giving your neighbors the heads up about additional traffic and noise coming from your property is a nice thing to do. It allows your neighbors the opportunity to adjust their schedule to accommodate the working being done. Heck, maybe they will let you know they were thinking about getting some roofing work too and you guys can work out a combo deal! At the very least, you will look like a courteous neighbor, at least until you throw that next loud party to show off your roof to your friends.

Figure Out What Exits You May Have to Use During the Roofing Work

ONCE THE ROOF installers have arrived, ask them what doors will be accessible for entering and exiting the house. Leave tape or posted notes on the inside and outside of the other doors during the work to help remind less attentive family members not to use those doors. While you're at it, communicate with your roofing company before, during and after the process to help ensure that everyone agrees, and the work goes as smoothly as possible. Asking how to install a new metal roof versus a simple roof shingles install might open your eyes to work and materials needed.

Final Words on Your New Roof

ARMED WITH THESE SUGGESTIONS, you have created the easiest imaginable situation for both your roofer and your family. If you take care of the pre-roofing checklist above, you have done everything in your power to allow the work to be completed quickly and painlessly. How long a roofing install takes depends on the skills and speed of the roofers you hired, but if you have done the above and have an average size house in most cases, they can complete it within 3-4 days at the latest.

the best energy efficient roofing

. . .

What You Need to Know About the Different Types of Roofing Materials and Roof Products

HUMANS NEED three major things for their survival. These are food, water and shelter. In our modern times, it has become easier for us to get these three necessities, but we need to ensure that we build our shelters to last. We want to make sure that the home that we worked hard to get can last with us a lifetime. Sometimes over one lifetime, if we plan on passing it down to our children and loved ones. This means it is more important now than ever to have the proper maintenance done on your home. So if you're curious to know what the best energy efficient roofing looks like, read on.

ONE OF THE most overlooked areas of the home can be the roof. It makes sense since it is the only part of the home we cannot easily see daily. Yet ironically, this is the most important part of your shelter as it is the first line of defense from what we needed shelter for in the first place — the elements. More especially the weather. It comes in a variety of flavors, but all of them can damage to your home, even that nice sunny day.

. . .

THOSE THAT LIVE in the northeast are all too familiar with varying weather. We have a humid continental climate, which translates to four unique seasons and weather, having a wide range throughout. There's everything from a hot summer's day to a horrendous snowstorm that comes out of nowhere.

IT IS a tough choice when selecting the best roofing. Some roofing options are better for certain conditions and could be detrimental to others. While this is not an exhaustive list of all roofing types and materials, it is a guide to the basic choices available for the average homeowner.

What Elements Damage a Roof and How Can I Ensure Longevity for My Roofing System?

The helmet of our shelter takes quite a beating all year long. From cosmic rays hitting the roof invisibly to snow and hail during the winter months. Let's look at the common weather North American homeowners face and what it can do to your roof.

Rain

THOSE OUT OF blue rainstorms we often experience during the end of summer and spring are pelting your roof in a constant drumming. Left unchecked, those raindrops can eventually find their way through the roof and cause leaks within your house. Your roof needs to be checked at least once a year to ensure that this issue is prevented. A newer roof has not had time to expand and contract due to heat and cold and therefore is less likely to have any major or minor cracks.

Snow

UNLESS IT'S CHRISTMAS MORNING, nothings worse than a snowstorm. As we all know, if the meteorologist calls for 2 inches of snow, we might see 2 feet and if they declare that a horrible storm is on the way and the kids are called off school, it's usually just a light dusting of snow. No matter what, snow means more than just a sore shoveling back and slow drivers. During our annual snowstorms, two major roofing issues happen here. The first one is that it will add weight on your roof that will cause structural damage (you do not want any weight on your roof, except for the material of the roof itself). The other issue that can happen is ice dams that will cause the snow to melt on the top, run down to the edges of your home, and then refreeze into ice, which will slowly seep into your roof deck. That refreeze is what will cause those teeny tiny roof cracks to eventually become giant gaping holes that require professional roof repair or, sometimes, a total roof replacement.

Hail

THIS WILL HAVE an immediate effect as it is ice balls hitting your roofing. They can cause dents or cracks in your roofing (depending on the material) and, after every hailstorm, you must inspect the roof to see the damage that was caused. A strong hailstorm can be especially problematic for metal roofs. When combined with a metal roof, hail can cause anything from cosmetic damage to the appearance all the way to real structural damage at the seams and joints of the roof.

What Can You Do to Protect Your Roof?

You can start by conducting annual or semi-annual roofing inspections. This allows you to spot potential issues and hazards before they grow into more expensive problems. Next is ensuring that no matter what happens, you have the best quality roofing supplies and materials protecting the top of your home. Not only do you want to protect yourself from the natural elements, but you also want to look for roofing that is energy efficient. Especially when every penny matters, there's nothing worse than having a roof that doesn't properly insulate or ventilate well, which means you'll spend more on heating and cooling. Let's look at some of the best energy-efficient roof materials for any budget. Spending a little more for an energy efficient roof might end up saving you a significant amount of money over the long run. The better the decision, the less likely you will ever need to type "roofing repair near me" into your search bar again.

Are Metal Roofing Suppliers the Answer?

When someone says metal roofing, what comes to mind? For many people, they think of the old ugly metal roofs of years past. Thankfully, in the past few decades, metal roofing has come a long way in both esthetics and function. Metal has exploded in popularity these last few years and for good reason. If it works for tanks it should work for your home, right? Metal roofing is an excellent choice when it comes to picking the right roofing material for several key factors. Of the roofing materials we will look at today, metal roofing is certainly the lightest of the bunch, which is great when the snow piles on. They also are fireproof because of the material and have a naturally reflective surface to push those nasty sun rays away.

. . .

THERE ARE some things to consider. They are more prone to denting (from hail and debris) and can actually contract and expand because of severe hot or cold weather. In addition, on the off chance this metallic surface needs replacement, the sections are much larger than other types of roofing, so it can sometimes be more costly.

How Long Can a Metal Roof Last?

A well-constructed metal roof can last anywhere from 40 to 70 years. That is certainly longer than most other roofing materials, but it's important to keep in mind that a metal roof usually costs a bit more. A metal roof supplier will give you specific pros and cons to their particular roofing system, which should give you a better idea of if this is the right roofing type for you. When a metal roof is installed most times, they can construct it right over existing asphalt shingles. This will save you significant time and money, not to mention the environmental impact of sending all those old shingles to the dump.

How Energy Efficient Is a Metal Roof?

Metal roofing can be extremely energy efficient, but it depends on how it is installed. An unpainted metal roof will naturally provide decent solar reflection, which will help keep the upper levels of your house cooler. By painting a metal roof, it can further increase its ability to reflect the sun's rays and increase its thermal emittance rating. Meaning by simply painting a roof with energy efficient paint, you can dramatically decrease the cost of energy usage during the warmest months. A metal roof that is hot to the touch has a low thermal emittance, and a properly painted metal roof with a high thermal emittance is cooler to the touch, meaning it is radiating that heat back outside rather than into your home. Asking your contractor about the best reflective coatings for roofs and specifically the best metal roof coating, could save you thousands over the course of its lifetime.

Are Asphalt Roof Shingles a Good Choice for Energy Efficiency?

If you're looking for one of the cheapest yet durable options out there, look no further than these types of roof shingles. Be warned, though, that you need to ensure that these are energy efficient from the get-go. Asphalt roof shingles, that are not rated as efficient, can actually be some of the least efficient materials for roofing out there. Check the labeling and ask the construction team you might be using. If energy efficiency is a major concern of yours, and it should be, discuss this with your roofing contractor up front so they know to use the type of material that will be most efficient. Discuss the differences in asphalt shingles (pros, cons, cost, and efficiency) before committing to any work.

SOME ASPHALT SHINGLES come with a specialized coating on them to prevent damage caused by Mother Nature and improve fire resistance. All asphalt shingles have the added benefit of providing increased sound insulation, in case you want to blast your music and not bother the neighbors. They also can come in a wide variety of shapes and colors, making it easy to provide something aesthetically pleasing, while also easy to maintain and replace.

How Long Will My Asphalt Shingle Roof Last?

WHEN CORRECTLY INSTALLED and properly vented with roofing vents, you can expect a new asphalt shingle roof to last at least 20 years. And considering that the cost to replace these types of roofs is considerably lower than other roofing materials, when judging strictly from a pricing viewpoint, it's tough to beat this type of roofing material. Most basic manufacturer's warranty will cover the roofing material for between 25-30 years, though some have much longer warranties. Just be sure to ask for energy efficient roof shingles as well as ask about the most energy efficient roof color.

What Are the Pros and Cons of Tile Roofing?

If you want to go for that beautiful look with some of the best durability out there, look at Tile Roofing. If you plan on staying at your home and eventually passing it down to your family, take the time to think about investing in such a long-lasting material.

Pros of Tile Roofing and How Long Will a Tile Roof Last?

It is not unheard of for a tile roof to last 50 years or more, if well maintained. As long as you do not have to worry about fallen branches cracking the tiles, they are long lasting and require little maintenance at all. It's a good idea to check the roof a few times per year to ensure that no tiles have developed cracks that could allow water to penetrate the surface. In addition, because of their unique shape, they might be one of the best choices out there from an energy efficiency standpoint. Tile roofs have fantastic circulation, so energy can be conserved year-round. The spacing underneath the tiles offers a place for hot air to escape, keeping it from the attic and upper floors of your home. We usually find them in that light brown/maroon color, but also have the distinct advantage of coming in multiple colors and shapes. They are also extremely durable against most damage that weather can cause as well as fireproof.

Cons of Tile Roofing

Tile roofing, though great for some, like anything else, has some major drawbacks as well. First is that it is a very heavy roofing material. Remember earlier when we discussed the impact that weight has on a roof? Well, tile roofing is among one of the heaviest and so special care must be taken to ensure the structure of your home can support this added weight, plus any weight from snow or ice that may accumulate in winter. Also tile roofing is not cheap, often it can cost almost twice that of asphalt shingles. Last, while typically very durable, when it is hit by tree branches or debris, it can become very brittle as they typically make it from clay or concrete.

What is Green Roofing?

Though rarely considered when discussing a roof, green roofing has become more popular in recent years and is particularly suited to city environments. If you have a flat roof, you might consider covering it with a garden or even basic plant covering to help with energy costs, insulation, stormwater management. This type of roofing is much more expensive and requires the help of a professional who's trained in sustainable green roofing for installation. Simple rubber coating for roofs combined with green roofing can be an effective way to get both longevity and lower your monthly energy costs at the same time. An experienced roofer who has built green roofs in the past will work with you to carefully assess your property to decide if this is a viable option or if you should consider a different roofing material.

What About Solar Panels, Do They Work Well?

If you want to go even further with energy efficiency, consider Solar Panels. Yes, even in inclement weather with snow and rain, there have been significant advancements made to increase the durability of these energy creators. Yet again, because of the wide variations in the weather, it would be best to ensure high-quality roofing is placed first. The best roofing company will find the right balance. When considering solar panels, know that they are also not all made alike. It is important to focus on those especially for inclement weather so that they can maintain their efficiencies during those bright sunny periods.

In fact, solar companies have improved the technology to such a degree that most homes can actually save money on electric each month and have the entire solar panel system paid off within 10 years or fewer. And unlike the solar power of years ago, today's solar panels are usually stronger than standard roofing tiles. So even though PA does not receive the same type of sun that a state like Arizona or California receives you can still feel confident that the panels will last through the abuse of our different seasons and generate energy day after day eventually helping the roof pay for itself.

TODAY YOU CAN GET solar panels for no money out of pocket with a financing plan built to not only save you money on your electric bill each month, but also build equity in your home.

Houses with solar sell for more than those without it. In addition they sell faster, and in a competitive market that can make a huge difference. If your house receives a fair bit of sun, it's almost a no brainer as it allows you to save money monthly and current government incentives make the switch to renewable clean energy even cheaper.

Final Words About Installing an Energy Efficient Roof

It can be tough not knowing how to approach that roofing, and it is also important to note that you cannot place roofing materials if the weather is too cold (around 40 degrees Fahrenheit) or too hot (around 90 degrees Fahrenheit), otherwise the tiles won't settle and adhere properly. This means that this is something that cannot be rushed and must take the time to make the best decision. Keep in mind the budget for the roofing and the energy efficiency it carries as it relates to the rest of the household. The first step is to get written estimates from local roofing companies as they will know what works best in your area.

will a new roof reduce your insurance premiums?

. . .

How a New Roof Can Lower Insurance Premiums and Other Tips to Lower Homeowners Insurance

NO ONE WANTS to pay higher insurance premiums than they have to. Whether it comes to your car, home, or life insurance, we are all looking for the highest quality insurance at the lowest possible price. It's not that we are cheap either, it's just that the money you save on homeowner's insurance might just be enough to upgrade your deck or afford that premium fire pit ring. Let's discuss how a new roof might in fact lower those annual premiums and other ways for the frugal homeowner to save on their homeowner's insurance.

Will a New Roof Reduce My Insurance Premium?

THE CONDITION and performance of your roof affects the premiums. Think about it, it saves money for insurance companies by not having to pay for the damage that can happen to your home if your new roof has better technology incorporated to protect it. It might sound funny hearing the words technology and roofing in the same sentence, but over the past few decades there really has been an enormous advance in the science behind roofing materials. If you have a new roof installed, it may help to lower your insurance rates and decrease your chances of incurring damage from natural disasters. Safety precautions play a role in lower insurance rates. There are many types of roofs and each has its own pros and cons.

ON AVERAGE, a new roof will have the following lifespans barring any natural disaster:

- **Asphalt Shingles: 20-30 years**
- **Concrete or Clay Tiles: 35-50 years**
- **Metal Roofing: 30-50 years**
- **Rubber Roofing: 30-40 years**

OBVIOUSLY, when a new roof is constructed, your insurance adjuster will consider the type of roofing installed and the expected lifespan of its components. Before the installation of a new roof, you can ask your trusted insurance agent what they would recommend to reduce the cost of premiums as it pertains to roofing materials. Their answers may surprise you, because while some roofing materials are indeed expected to last longer, they may also be significantly more costly to repair. In addition, even within the particular category of roofing material you choose, i.e. asphalt shingles or metal roofing, there are various types of each. You might find an asphalt shingle with advanced coatings to better protect it from algae growth or make it more storm resistant. Those might qualify you for additional savings from your insurer. When thinking about a new roof, be sure to research any possible insurance breaks by discussing it with both your roofer and then your insurance agent. Often homeowners policies can be reduced by 5% or more, which could translate to thousands in savings over the course of the roof's lifespan.

Why Do You Need to Update Your Insurance Information When Getting a New Roof?

BESIDES THE OBVIOUS OF SAVINGS, even minor roofing repairs and maintenance help to keep your insurance rates low. Maybe you added additional roofing vents to better expel the hot air in your attic. That is something that might be worth telling your agent in case they can note that in your policy and find a reduction. For instance, if you upgrade from an asphalt shingle roof to lets say a metal roof, it might surprise you to know that the insurance company considers one roof to be significantly more resistant to both fire and rot. That, of course, translates to less expense incurred by the insurance company should the unfortunate occur. It's best practice to contact your insurance carrier whenever a major change is made to your home to ensure that your coverage aligns with the value of the home and its contents.

How Much Will Your Insurance Premium Change After You Install a New Roof?

THERE IS no direct answer to how much they can adjust your premiums after the installation of a new roof, but it's possible to see an annual reduction of anywhere from 5% to 25%. When factoring in the cost of a new roof, take that savings into account and it may make a more expensive roof installation cheaper in the long run than a simple roof repair.

Which Type of Roof Does Most People Have for Their Homes in North America?

THE MOST COMMON type of roofing material can vary from state to state. For instance, tile roofing is much more popular in Florida because of its long-lasting durability and increased heat dispersion. In America, by far the most prevalent roofing material is basic asphalt shingles. These basic shingles carry a significantly lower cost, fire-resistance, and hold their own in both hail and snow. That said, asphalt shingles are not necessarily the best choice for YOUR home. There are several factors you need to consider, such as the time you will live at your home, how each roofing material affects resale value, energy efficiency, insurance costs, longevity, and weather resistance. Talk over each factor when discussing the best choices with your roofing contractor before simply picking the lowest priced option.

What Are Some of the Best Ways to Reduce Homeowners Premium Rates?

HERE ARE some basic tips for further lowering your homeowners' premium rates before and after you consider a new roofing system:

• **SHOP AROUND** - Just because you have been with your insurance agent for years doesn't mean they are giving you the best rate. Insurance agents are people and while your home is top of your mind, it's probably not top of theirs. Check to see if they are offering any new savings products that might help your current rate. Let them know you are checking the rates of other carriers too and trying to find the best deal for you and your family. Next, visit this link and request free quotes from competing insurers to see the best deals. Sometimes, you might save thousands, but in the worst-case scenario, you will sleep better at night knowing whether you have the best price and coverage available.

- **INCREASE Your Deductible** - By simply increasing your deductible, which is the amount of money you pay before the insurance company steps in, you can save a significant amount without a reduction in overall coverage. This works for home and auto.

- **COMBINE INSURANCE PRODUCTS With a Single Company** - By including your home, auto, life, umbrella and any other insurance products into one company, you may find additional savings.

- **FOCUS ON REBUILD Costs** - The land your home sits on isn't at risk from most perils that your dwelling is. Make sure you are focusing on rebuilding cost rather than the overall amount you paid for your home which usually included the land underneath it.

- **CONSIDER RETROFITTING** - If you have an older home, upgrading electrical or plumbing systems while costly might help prevent major catastrophes from occurring in the future. Much like a new roof will lower the premium from an aged roof, a new electrical or plumbing system can do the same.

. . .

- **BETTER SECURITY** - Every house has basic smoke detectors (or at least it should), but that's simply the lowest rung in home safety. Many newer homes have advanced burglar alarms that call the police, sprinkler systems, or advanced fire detection. These features are valuable to an insurance company and should lower your rates considerably.

- **MAINTAIN OR FIX your credit** - Good credit is essential these days and if your credit has slipped, it might be time to work to rebuild it. Modern insurance companies are using credit scores more and more these days to determine the insurance costs you pay.

Final Words on Insurance Premiums

WHILE SPENDING the afternoon discussing ways to lower premiums may not sound as sexy as sitting on the beach, it's a better use of time than almost anything else you could be doing. Savings you earn from shopping around or talking to your agent will continue to save you money month after month and year after year for life. Upgrading your roof and finding other ways to make your home less of a risk protects you and protects your insurance costs.

WHILE I DO NOT GUARANTEE insurance savings when you replace an aging roof, it is ALWAYS worth it to spend the time to see if you may save money. Worst-case scenario is you catchup with your insurance agent and they can tell you about any new products or discounts available to you. There are worse ways to spend an hour than potentially saving money every single year.

spring cleaning roof checklist

. . .

SPRING IS the time to get your home ready for that hot summer weather. One area that's easy to overlook is your roof. Cleaning a roof is important because of the many issues that can arise if it's neglected, especially after a heavy winter. Not only that, but cleaning the roof makes the house more aesthetically pleasing, and it could even increase its value if you plan on selling it.

IF YOU'RE NOT sure where to begin, here's a checklist of the five most crucial actions to help you get started.

Clear out the gutters

CLEAN OUT THE debris that washed down from the roof during winter storms. Make sure the downspouts are also clear and draining properly, so rainwater will flow away from your foundation and not puddle around your house.

Clean up around your skylights and chimney

Look at the flashing around these features and make sure it's still intact and not letting water in where it doesn't belong.

Trim tree branches that hang over or near your roof

If branches hanging over the roof get weighed down with snow, they can break off and damage shingles or gutters when they fall. Also monitor branches that grow close to power lines that lead to your house — if they get tangled with those lines during a windstorm, your home could be at risk.

Clean off your shingle

After winter has taken its toll on your roof, it's a good idea to give it a good clean off. A roof cleaning product can help wash away dirt and improve the overall look. Later in this chapter I will tell you how to make your own roof cleaning solution at home for cheaper than the store-bought chemicals.

Check for leaks

Go into your attic armed with flashlights and check for signs of moisture along every rafter and truss. Make sure you check any areas that have had previous water damage as well. If you find any leaks, call a professional roofing company to repair the damage as soon as possible.

Why Bother Cleaning Your Roof?

OUT OF SIGHT, out of mind, right? You may think that you don't need to clean your roof because it's way up there and no one really sees it, anyway. But what you might not realize is that while your roof might be invisible from the ground, it could have endured some damage from seasons changing and built up debris. Roof cleaning is one of the preventable items on the honey do list that has multiple benefits:

Healthier Environment

THE FUNGI, algae, and moss that grows on your roof can damage its structure and lead to leaks. This will cause rainwater to collect in places where it isn't supposed to, leading to mold growth, which could be very hazardous for your health. Cleaning your roof will keep these pathogens from growing out of control and causing problems.

Safer Home

A CLEAN ROOF also means a safer home for you and your family. If there are shingles missing or damaged, then they could fall off at any moment if the wind is strong enough. And if you have debris on top of your house, then this could fall down on people or pets below.

Curb Appeal

ANOTHER BENEFIT that homeowners sometimes overlook is that a clean roof improves curb appeal. If you are planning to sell your home soon, a clean roof makes a big difference. Real estate agents, stagers, and homebuyers all note the condition of the roof!

Prevent Plant Damage

Moss, lichen, algae, and fungi can cause serious damage to your shingles over time by eating away at their protective granules. If there are branches overhanging your roof, they can scrape away the granules when they blow around in high winds or during storms. Tree sap can also cause staining on roofs that is difficult to remove. Keeping your roof free from these damaging elements will help extend its life considerably and prevent costly repairs down the line.

Prevent Heat Damage

A DIRTY ROOF is going to get damaged by UV rays. This is because UV rays are more likely to degrade roofing materials if there are dirt particles covering the protective chemicals that reflect sun and heat. That means that your roof will not have as long a lifespan as it should because those little dirt particles are going to speed up the degradation process.

Moss Growth

THERE ARE shady areas on your roof where Moss thrives. Moss holds moisture against your roof, promotes decay of the shingles, and weighs down your roof. This last point is most important because extra weight on your roof can lead to water infiltration and damage to your home's interior.

Gutter Blockage

DEBRIS BUILDUP in your gutters prevents proper drainage and can cause damage to soffits, fascia boards, and window sills. It can also be a breeding ground for pests of the flying and slithering variety.

Why You Might Hire a Pro

ROOF CLEANING IS DANGEROUS. Many roofs are steep, and hard to access. It's easy to slip or lose balance. A professional roof cleaner has the knowledge and experience to do the job safely. They also have the right safety equipment, like harnesses, safety lines, ladders, and other specialized gear to get the job done safely. And it's not just safety equipment, something as simple as a gutter guard is common practice for a pro, but maybe not top of mind for a DIYer.

Equipment

PROFESSIONAL ROOF CLEANERS have access to advanced equipment and chemicals that ensure maximum impact with minimal risk of damage. Commercial-grade power washers remove mold, mildew, and other stains from your shingles without harming them. They also use advanced solutions that kill mold and mildew but don't damage your shingles or vegetation around your home or business. When you visit your local hardware store and purchase what's on the shelf you may find yourself taking an additional trip later in the season, because the chemicals didn't have the lasting effect that commercial-grade does.

Better Results

A PROFESSIONAL HAS the expertise and training to ensure that your roof is thoroughly cleaned of mold, mildew, and other stains. Cleaning products used by professionals kill 99% of all mold spores before they start keeping you safe from the elements and prevent your house from wasting energy. This is more than just a cosmetic issue. A dirty roof can be a breeding ground for mold, mildew, and hazardous materials.

Why I Would Do It Myself

ANYONE WHO KNOWS me knows I'm a pretty frugal guy and so when it comes to cleaning my roof, I would rather risk injury and death then spend more than I have too. If you're of a similar mindset and have a shingle roof, let me show you what you need to do the job right the first time.

Soft Washing Method

WHAT YOU NEED:

- Metal adjustable ladder
- Gutter Guard (Order Online) - Trust me this matters if you don't want to damage the gutters
- Three gallon hand pump to spray the roof - This is slow, steady and safe for you and the roof
- Small plastic gutter cleaning tool or cut an old half gallon milk jug
- Multiple 5 gallon buckets
- Dawn dish soap
- Pool shock chemicals

ONCE YOU HAVE ALL those tool its time to get to work:

1. Mix 1 gallon pool shock + 2 gallons H2O + 4 oz. dish soap (I prefer dawn) give it a good stir
2. Secure the ladder and put gutter protection in place
3. Clean the gutter of any debris and buildup so the solution can flow freely down the gutter
4. Gently spray your entire roof, but do not use a broom unless its very soft bristles and you absolutely must

THIS METHOD IS CALLED the soft washing and here is why it's my preferred method to clean a dirty roof. First, it's the only method recommended by the ARMA (Asphalt Roof Manufacturers Association). If you were to use a strong broom or wire brush you more than likely would be scrapping away significant shingle grit along with the mold and algae buildup and could shorten the life of your roof by years because of it. Depending on who installed your original roofing, using a wire brush or hard bristle broom might even void the warranty all together. The downside of soft washing is that it can take months for the mold and algae to finally fall off your roof since you're depending on the weather to push it off once you killed it with the chemical solution you made. So if you are looking for immediate results call the pro and make sure the can guarantee they won't void existing roofing warranty. One final note: **DO NOT USE A PRESSURE WASHER ON YOUR ROOF**, using one could cause massive damage to your roof and also remove shingle grit again shortening the roofs life. Most shingle warranties will void if you use a pressure washer, so don't risk it.

MAINTAINING your roof is not only good for your home, but it's also good for the environment. Plus, it's easy. If you don't have the time or knowledge to climb up on your roof and clean it yourself, just hire a professional to do all the heavy lifting for you. Either way, make sure you add the roof to your spring cleaning list.

how do i find a good roofing contractor?

. . .

SO YOU NEED to find a good roofing contractor and you don't know where to start? Well, an independent roofing service is just a call away, right? Shouldn't there be some sort of roofing contractor list or directory that has them all ranked 1-100 with the cheapest best quality ones right up top so you can save money and have a perfect experience?

THE PROBLEM IS, with hiring a reputable installer for a roof replacement, there is no perfect database. Finding a roofing contractor ratings list along with good weather guarantees, is as likely as tooth fairy at a dental cleaning. But if you're looking for a step-by-step guide to take you from start to finish and ensure you and your family have the smoothest, most stress-free experience possible, we got you covered.

WE GET IT, finding a professional roofing contractor is scary, but whether you want to get multiple top quality roofing contractors competing for your business, or you just want to check and ensure they're legit this chapter will help you make a better informed decision.

We can't promise it's always fun, but when it's all finished and you have a beautiful new roof that cost you less than you expected, you might just think it was worth it!

Find a Local Licensed Roofer

This can be done a few different ways. You can begin by visiting The National Association of State Contractors Licensing Agencies (NASCLA) which maintains a list of licensing boards. Go to the NASCLA website and look under "Certified Contractors" for the state in which you live. If none are available, another way to find a licensed roofer can be done by visiting the Department of Consumer Protection and State Attorney General in step 2.

Find a Roofer That Has Verified Their License by Signing the Contractor License Agreement

Next step is verifying the License and Contractor Agreement along with checking the roofing company's certification status. You can also learn lots of valuable information to help you make an informed decision regarding your roofing repair or replacement. This includes information regarding The Home Improvement Consumer Protection Act, which safeguards you from unscrupulous contractors and illegal advertising claims. The most important step is to make sure the roofer holds a current certification or is registered within the state.

Check the Better Business Bureau for Any Registered Complaints About Your Potential Roofing Contractor

The Better Business Bureau has a vast database that contains most, if not every, legitimate roofing contractor within the state. You can type in your area and find a list of local roofing contractors that are operating in your county. It also allows you to search only for BBB accredited contractors, which are a select group of roofing contractors and roofing companies that have gone through careful vetting and evaluation. To become accredited, this group has to uphold all BBB standards and have a track record of doing so. While it is unnecessary for you to hire someone who is accredited or even listed on the BBB, it adds a level of security and safety to the enormous investment you are making.

https://www.bbb.org/

Use the Resources From GAF Materials Corporation

Another tool is to visit GAF.com, which is one of the largest roofing material manufacturers in the US. They perform an annual certification of roofing contractors near you and check them for valid commercial insurance and years in business before granting a certification. While it may seem that this might be the easiest method of finding a contractor, remember that they are in the business of selling materials to roofers and not all roofers use what they create and might therefore find little value in associating with their brand.

. . .

https://www.gaf.com/en-us/for-homeowners/how-to-find-a-contractor

Visit the National Roofing Contractors Association Website

Besides visiting the Better Business Bureau and GAF, one last resource in your quest for a top quality roofer is to visit the National Roofing Contractors Association page and locate your state member directory. This will provide you with the smallest list of contractors. While it's great to ensure the quality of roofing contractors who will work on your home, it's important to remember that just because someone isn't a member of the NRCA doesn't mean they are not a great contractor and more than qualified to perform your roofing repair or replacement. In the same way, just because they have an excellent reputation within the BBB, that does not mean your due diligence is completed.

https://www.nrca.net/Members

Get Local Referrals From Family, Friends and Neighbors

The most important tool in your tool belt is a referral from someone you know and trust who has "recently" had roofing work completed. It's important to ask when the work was completed last because contractors and companies can change from year to year. The high quality roofing shingles they used in the past might have been downgraded to a cheaper brand of shingles with no warranty. Likewise, the cheap roofing contractor who was just getting started and offered the lowest price roof repair may have studied their craft and now charge accordingly. Ask them questions like:

- **Why did they choose that roofing company?**
- **Would they use that contractor again? Why or why not?**
- **Was it the cheapest roof estimate they received?**
- **Did the quality match the expectation the roofer set?**
- **Did they stay within the budget for the roof repair or replacement?**
- **How did they first hear about that particular roofing company?**
- **Did they receive multiple quotes for the roofing work and did the contractor take the time to explain how the job would be completed?**
- **How long did the work take and how did they protect your roof when they were not at the job site in case of rain or inclement weather?**

The advice you receive might help you avoid an over-priced or under-qualified roofer. Perhaps, it might help you find a contractor who is both high quality and fairly priced. If they had a positive experience, chances are high that you might too.

Final Words

Finding a licensed roofing contractor can feel like a monumental undertaking. From that first Google search of "roofing repair contractor" till the moment of your first meeting and quote, it takes a little while, but it's essential. Likewise, there are many other factors to consider and specific questions YOU NEED TO ASK before the work begins. Not knowing exactly what to ask, a roofing contractor might end up costing you thousands on your roof repair or replacement. The next chapter will tell you the essential questions you need to ask before the work begins.

17 questions to ask a potential roofing contractor

. . .

What Do You Need to Look For When Hiring a Local Roofer

HERE IN NORTH AMERICA, we are lucky to have plenty of able-bodied hard workers willing to try their hand at your roofing project. No matter where you live, chances are there is someone just a few blocks away willing to work on your roof. Unfortunately, while many "roofing contractors" are quality professionals, some "professionals" are all too willing to make your roof their guinea pig. Don't worry, finding a quality roofing contractor that can help with your roofing project is easy if you know what questions to ask.

While it may be difficult to sort through potential contractors, there are a few questions to ask and the answers you're looking for to help you decide:

1. How Long Have You Been in Business?

A ROOFING CONTRACTOR is someone who installs, repairs and maintains roofs on homes or commercial buildings. Established roofing companies are doing something right in the sense that they've been in business for a long time. At minimum, you want to see a track record of at least a few years of experience under their belt. Obviously, the more experience, the better because it means you are unlikely to have some unique problem they have not come across before. That said, if the price is right and a younger contractor can convince you to take a chance, they might reward you with new innovations that they recently discovered in trade school or an apprenticeship. A less experienced contractor may work significantly cheaper because they lack the experience necessary to charge more. While saving money is always nice, if you face problems down the road because of poor craftsmanship, it might not be an even tradeoff.

2. Are You Licensed?

NOT ALL STATES required contractors be licensed, however many municipalities still have their own necessary licenses needed before picking up a hammer. Many are required to register with the Attorney General and you can visit your state's Attorney General and search for information about contractors within the state. Remember, these government officials are here to work for you and make your life easier, so use the resources they provide.

3. Do You Have Workman's Comp Insurance?

THIS PROTECTS you and your home from injuries or damage resulting from an accident at work. THIS IS IMPORTANT!

4. Do You Carry General Liability Insurance?

SEE ABOVE.

5. Who Is Your Insurer?

VERIFY who their insurance is through and, if possible, have them show you their coverages.

6. Is Removal of My Old Roof Included in Your Estimate?

IN MOST CASES, except for metal roofing, they should always remove old roofs before they put new down. You would hate to be surprised by an additional bill for garbage removal after agreeing to terms with the roofer. Make sure they include this or ensure you have factored in the price if it's not. These days, most roofers will include this in the contract and the few that don't should be able to give you solid recommendations.

7. Will You Be Installing Edge Metal or Drip Edge When the New Roof Is Installed?

THIS IS a small piece of metal that extends to the gutter and will protect your roof from problems down the road. You might think every roofing contractor would obviously add this for your benefit, but unfortunately, this isn't always the case. Ensure that the roofer you hire will add the additional drip edge or edge metal to protect your investment.

8. What Do You Use to Protect My Gutters From Your Ladders During Installation?

Most roofing contractors will use ladder stabilizers to protect your gutters and avoid putting unnecessary weight and pressure on them. If they don't plan to use ladder, stabilizers, ask them what method they are using to protect your gutters from damage during the roofing job.

9. How Do You Dispose of Refuse and Debris? How Do You Ensure No Nails or Old Roofing Is Left Anywhere in My Yard?

This will tell you how they plan to get rid of all the trash from the job. Hopefully, there is a quick way to eliminate it, rather than a large dumpster sitting in your driveway for weeks till the trash company comes to haul it away. It is also a good idea to remind them of small children and pets to make sure they are extra attentive to making sure every tiny nail and small piece of roofing wood has been removed. Still, after the job, you will want to do a thorough inspection yourself along the sides of your house and check for anything that they missed.

10. What Happens to My Roof if It Rains During the Job?

If they say anything other than that, they will be carefully covering your roof with a tarp or plastic and checking to make sure no water is leaking through find a new roofer.

11. Do You Have a Local Telephone Number and Address?

While not as big of an issue in the northeast, during hurricane season in the southern part of our country, roofers from all over flood the area and might not be as familiar with the intricacies of that neighborhood's roofing issues. While you are most likely dealing with someone local, it's still nice to have a number and an address to ensure you can reach them again should an issue arise.

12. What Is My Warranty on the New Roof and What Does That Cover?

Over the past few decades, roofing technology has come a long way. Most shingle roofing should have a 25 year warranty and be able to make it at least that long without a replacement. Other types of roofing may not come with the same warranty such as metal roofing, which typically only has a warranty associated with the paint that is used on the metal, but a warranty can usually be negotiated. Ask lots of questions and get all warranty information in writing.

13. What Is the Cost of Plywood Should You Find the Roof Decking to Be Soft or Rotten?

Make sure they are giving you specific information regarding the cost of new plywood to replace soft or rotten wood they remove. This will ensure that they do not overcharge you for plywood, which, after the job is complete, can be difficult to dispute.

14. How Do You Protect My Landscaping During the Roofing Job?

DEPENDING on the level of pride you take in your landscaping, this can be an issue. If the landscaping matters to you, make sure you communicate that and let the roofers know to protect your plants and bushes from falling debris. Find out if they will replace damaged plants. That usually helps ensure they are extra careful during the job.

15. Do You Provide Written Estimates Detailing Scope of Roofing Work?

YOU WANT A WRITTEN ESTIMATE. This should detail all associated costs (trash removal, cost of new plywood, cost of the new roof, etc.). Don't settle for a verbal estimate. It must be written and explained clearly to you so you understand it. Go through it WITH the contractor item by item and ask for clarification for parts that seem vague or you don't understand.

16. Do You Have References and/or Pictures of Previous Roofing Work?

IF YOU ARE DEALING with an experienced professional, most should have pictures of previous work and references who can speak to their skill and competence. If they have neither of those and preferably both, consider it a pretty large red flag.

17. How Soon Can You Begin the Work Once We Agree on the Details?

IF THIS IS something you need done right away, make sure that your schedules are aligned. Keep in mind that often the area's best contractors might have many homeowners trying to hire them, so if time is of the essence it might cost more or you might have to settle for a lesser known commodity.

Final Words Before You Begin Your Roofing Project

LOADED WITH THESE 17 QUESTIONS, you can be sure that you are dealing with a qualified professional and not some Johnny on the spot looking for a quick payday. If your contractor is vague or unwilling to answer any of the above questions, that might mean its time to look somewhere else to get the job done.

REALTORS WILL TELL you that a quality roof will not only increase the curb appeal of your home, but will increase the property value along with it! Consumers understand the importance of a good roof because of the dramatic changes in weather that we all experience. Don't let an old roof detract from an otherwise safe and beautiful home. Remember, cheaper isn't always better! It's best to pay a fair price for quality work to ensure you don't run into trouble down the road.

how often should you clean your gutters?
. . .
The Benefits, Why It Matters, and if You Might Need a Professional Gutter Cleaner

What Is the Importance of Cleaning Your Gutters?

MAINTENANCE in our lives is very important. We need to maintain our health, maintain our families, and maintain our homes. This seems like a no brainer but always seems to be overlooked. Since we, as humans, use our senses primarily, we don't always see an issue until sometimes too late. Ensuring our home is maintained is also no easy feat. There are so many moving parts, we're just happy when the walls stay up and the heat stays in during the winter.

HOWEVER, what often happens is that we forget a critical component of our home, which is our gutters on the roof. Not everyone thinks these are important, but if you are in an area that gets a lot of rainfall (and we all know Pennsylvania does), then these are critical. They help provide a route to collect and drain rainwater off your roof, and into designated places. This helps reduce water and moisture build-up and ensures longevity with your home components.

. . .

For most of the US, we get all kinds of weather, from rain to snow to hail, too hot and humid summers. As the seasons shift, our roofs accumulate a lot of items via the gutters, which are the arteries of the house and are needed for proper drainage. This accumulation of nature is affectionately referred to as gunk.

We're going to cover several topics related to how often, why, and how to clean our gutters. Keep in mind that there are several tools out there, including a gutter guard that could potential reduce how much cleaning needs to be done. But even with these tools in place, you still need to check twice a year and clean your gutters as needed.

Let's get one thing out of the way first. With preventative maintenance for your home, hour for hour, there is no more important task that a homeowner could be doing. Gutters play an oversized role in the protecting the value of the home we live in. While it may not seem like a big deal, clogged gutters can cause huge (read expensive) problems and compromise the structural integrity of your home.

Three Reasons You Need to Clean Your Clogged Gutters Yesterday!

It's just rainwater coming down, right? Mother nature's life source. Wouldn't I just use water to clean water? It makes no sense, right? Mother Nature is beautiful, but also very messy and dirty. This is not filtered water hitting your roof and entering your gutters. This is a mix of rainfall, leaves, bugs, dirt and manmade materials that might come off the roofs, all to form that gunk that will clog your gutters faster than a double cheeseburger can clog one's arteries.

- **Clogged gutters will damage your roof, leading to costly roof repairs and replacements much sooner than otherwise necessary.**
- **Gutters filled with debris are an ideal breeding ground and home for rats, snakes, insects and all the rest of our least favorite variety of God's creatures.**
- **A clogged gutter system causes the water to flow incorrectly and instead of leading towards the yard or the street, it could lead directly to the home's foundation.**

Let's dive into each to give a better understanding of what's at stake with clogged, nasty gutters.

Clogged Gutters Will Damage Your Roof, Leading to Costly Roof Repairs and Replacements Much Sooner Than Otherwise Necessary

Your roof is all that's standing between you and the variety of elements we have here in the US. From the blistering summer sun to the snowy blizzard conditions of the winter, our roof has one job; keeping us comfortable and safe. Gutters play a huge role in protecting the integrity of the roof, as their sole job is to divert the rainwater away from the places it can do the most and most expensive damage. One of those critical places is the roof, which when water can penetrate its upper surface will quickly rot and disintegrate, causing to water and moisture in the places we least want it.

IF YOU ARE LIVING near a wooded area especially, it might always be a good idea after a major storm to check the status of your gutters. This is more of a status check than a need to clean, just to make sure that you don't need to move up the timetable for your cleaning.

Gutters Filled With Debris Is an Ideal Breeding Ground and Home for Rats, Snakes, and Insects

All that humidity and moisture is a bug's best friend. This is especially true for mosquitos and other similar bugs. Not only can dead pets get in the way and help form the ooze that will need to be cleaned up, but these gutters could also actually become a home to them. And they could literally start eating up your house and cause some serious damage.

SNAKES FIND their way into dirty gutter systems because of an abundance of things to eat, like mice, rats, small birds, and large insects. Mosquitos, which have even this year been found to be carrying West Nile Virus, make their home and breed in the moist areas of the gutter that pool water because of debris. If spending time outdoors is an activity your family enjoys, it will benefit you to maintain the integrity of a healthy gutter system.

A Clogged Gutter System Causes the Water to

Flow Incorrectly and Instead of Leading Towards the Yard or the Street, It Could Lead Directly to the Homes Foundation

This will lead to cracking in the stone, concrete, block and wood foundations as water penetrates in both freezing and thawing. Water is the number one cause of a cracked foundation, which can cost tens of thousands to fix. Of course, assuming structural foundation damage can even be fixed.

Ok – Thank You for the Scare – How Often Should I Clean These Gutters

A typical asphalt roof, while perhaps having a 25-30 year lifespan, will be dramatically cut short, necessitating replacement simply by not cleaning your gutters. Assuming you treat your body with better care than your house, a simple rule of thumb is to schedule a gutter cleaning at the same time you schedule your semi-annual tooth cleaning. That way twice per year leaves, debris, sticks, bird's nests, insects and whatever else finds its way to your gutters can be quickly and efficiently dealt with before an issue arises. This is the bare minimum. As mentioned above, depending on how bad a storm was, or how much wildlife is around you, it might require multiple check-ins throughout the year. We also recommend, based on the wide variety of weather in the Greater Pennsylvania area, that possibly upping the cleaning to quarterly. It's easier to clean a gutter than to have to remove clogs or replace them altogether.

How Can I Clean the Gutters Myself and When Is It Time to Call a Professional?

Gutters are easy to clean. The biggest issue is how, well frankly, disgusting it can be. That and how comfortable you feel being on a ladder for hours at a time. Unless you're an amateur handyman of any sort, it might be best to look at contacting your local roofing experts, as they will have the right roofing supplies and gutter cleaning equipment. While they are there on site, they can always do a quick assessment if all you need is just a cleaning, or if there is the potential of an underlying issue that is ready to spring up. Cleaning is not always a costly service, and if you hire a professional, you can stay clean and feel confident in a job well done. Where it gets expensive is if someone has not cleaned it regularly and there are issues that need to be dealt with to get your gutters back to tip-top shape.

If you fancy yourself as something of a handyman, or you like to get your hands dirty, and we mean really dirty, then find that old pair of Galoshes and overalls you bought once, and prepare to get into the muck. Wear full coverings and work gloves as well, because you might clean things you don't want to even think about. In addition, wear long sleeve clothing in case of contact with pests of the stinging or biting variety that have grown fond of their new home.

Now that you have made peace with the fact that you will do this yourself, make sure you have a sturdy ladder with a ladder stabilizer. This is a very important step because you do not want to accidentally damage the gutter. Without this stabilizer, the ladder will be propped against the gutter, which can push it down and break it. Let's focus on one issue at a time and ensure that we're using the right tools.

. . .

BESIDES THE LADDER, you need something to scoop up the gunk. Use your hands if you want to, but it is better if you use a tool designed for the job. Most hardware stores sell a gutter scooper that can be used to quickly and safely remove the gunk out of the gutters without damaging your gutters. Keep in mind that if you want to use something different, make sure it's made of plastic. It is all about preserving your gutters during the cleaning process.

PLACE some sort of tarp or something on the ground to pick up all the gunk and throw it away later. You do not want this all over your home in random piles of gunk. It can be unsanitary and also will not be aesthetically pleasing.

AFTER ALL, the major and obvious potential blockages are removed, take a garden hose and rinse the gutters as needed. If you did everything right, it should have the same effect as clean rain gently going down the length of your gutter system.

Doing a Spot Inspection After the Cleanup

After cleaning up the mess (and yourself), check to see if there are any dents or sagging in your gutter set-up. These can be easy to fix, but can also be included when using a local contractor who's doing the heavy lifting of cleaning, anyway.

AGAIN, although this is a completely doable DIY, gutter cleaning can be a lot of work. There's also a lot of strain on the body, as you're constantly standing on your feet and slowly moving down the gutter to clean it. Cleaning is also best done before winter, as all the gunk can convert into ice dams with the cold air. This will lead to water damage on your roofing and house as well as possibly collapse the gutter where the dam is.

REGARDLESS OF YOUR approach to this situation, gutter cleaning and maintenance is an often overlooked task that most people only notice when they see what is the equivalent of a beaver dam blocking their drainage. Do yourself, your family and your home a favor, and give it a proper gutter cleanse. Your house will thank you.

final words

...

I HOPE you enjoyed this book, and found enough information to help you save money and maintain a beautiful roof now and in the future. Think of this book as your North Star and should you ever need to consider a roof repair or replacement in the future be sure to utilize the checklist to ensure you receive quality work.

THANKS FOR READING AND PLEASE LEAVE A REVIEW

BEST OF LUCK!

What I Wish My Roofer Had Told Me

Stop Renting Your Energy
OWN IT JUST LIKE YOU OWN YOUR HOME...

Solar for $0 Down
Go solar for NO UPFRONT COST! Seriously, own your solar panels for ZERO down.

Increase Your Home Value
On average, solar increases a home's value by 4.1%, according to a new Zillow analysis.

Save Up to $200/mo. On Energy
The average electric bill has sky rocketed. Solar lets you lock in your savings.

Get a 30-Year Warranty
Protect your solar investment with a zero deductible 30-year warranty.

30% Federal Tax Credit
The Clean Energy Credit allows you to subtract 30% of solar costs off your federal taxes.

jonnelsen.com

bonus: solar powered energy theft preview

Chapters 1-3

READ the first 3 chapters from the companion book:

Solar Powered Energy Theft

- **Chapter 1: Better Energy**

- **Chapter 2: Panels + System**
- **Chapter 3: Is My House Right for Solar?**

better energy

...

Where Energy Comes From & Why Residential Solar Just Makes Sense

---

"Learning never exhausts the mind."
  - Leonardo da Vinci

WE FLIP on a light switch and instantly receive the benefits of electricity, but few people fully understand how that power makes it to our homes and how fragile that system truly is. Ask anyone living in Texas and they can recall when the system broke for weeks, leaving hundreds dead because of a power grid failure that left 4.5 million homes and businesses without energy or water. Moments like that are enough to cause anyone to worry about the future and what we can do to protect those we love.

The Current State of Energy in the US

IN THE UNITED STATES, the electric grid comprises an extensive network of independently owned and operated power plants and transmission lines. They created this system for supplying power from the point of generation to the homes, offices and businesses that depend on it to function often hundreds of miles away. They generate power in a variety of ways, but the common denominator is that it all still flows through the grid.

THERE ARE three main power grids in the U.S. and they are:
- **The Eastern Interconnect**
- **The Western Interconnect**
- **The Texas Interconnect**

YES, Texas has its own grid and in 2021, when it suffered several large winter storms during the month of February, that grid collapsed. As for the Eastern and Western grids, you find that those actually reach deep into Canada, helping to power the vast majority of our northern neighbor as well.

Where That Energy Comes From

WITH ENERGY PRODUCTION, the term "How the Sausage is Made" comes to mind. Most of the energy produced in North America comes from what we would consider dirty energy (coal, oil, gas, etc.). And while I have nothing against those industries or the people that work in them, much of the legislation and infrastructure surrounding them is based upon supporting the industry versus supporting the people.

. . .

OF THE VARIOUS forms of "dirty energy", coal is far and away the worst health wise for human, animals and the planet alike. In fact, the town of Evansville, Indiana, is within 30 miles of seven coal-fired power plants. These plants contribute to thick air pollution known to locals as "The Evansville Crud". Which is reported to leave a thick layer of particulates on everything from cars parked outside to children's swing sets.

THE REASON COAL is especially egregious to health and air quality is because it is a messy, inefficient system. When people think about pollution, they often think about the large smokestacks spewing toxins into the air, but it's actually much more than that. The extraction of coal itself, the machinery to dig it up and place it in trains and trucks, the transportation of it, and the long journey it has before becoming energy. Here is the process:

- **They extract coal from the earth.**
- **It's placed in a bunker for storage, then sent to a pulverizing mill**
- **Then several steps (bag filter, storage tank, feed tank, distributor) all with the potential for pollution before it even hits the blast furnace to generate the actual power and most of the air pollution associated with it.**

CONTRAST that with cities like Honolulu which leads the United States in per capita solar power and as a result has one of the highest air qualities in the U.S. Solar is not the only clean renewable energy source, but from a power perspective it is one of the best options. Other cleaner energy sources are wind, nuclear, and clean hydroelectric, which are all being built around the world at a rapid pace.

THE U.S. RELIED on these dirty forms of energy to power our growth, as many developing countries do today. Luckily, technology has caught up to industry and clean energy is now as cheap to produce as dirty energy. That is why, according to the federal government, in 2021, wind and solar supply 70 percent of the new power plant capacity being built. In addition, almost every new U.S. power plant of 2021 will be a carbon-free facility. This is great news as it means we are heading in the right direction to reduce pollution and our negative impact on the planet while also producing good paying jobs here at home.

How is Electricity Delivered?

Now that we know where the energy comes from, how does it make it the hundreds of miles to our homes every day?

1. **Power planet generates the electricity**
2. **Power flows to the transformer where it then steps up voltage for transmission**
3. **Power flows on transmission lines long distances until reaching neighborhood transformer**
4. **The neighborhood transformer steps down the voltage so we can use it in our homes**
5. **Distribution lines carry the electricity to our houses**
6. **Local transformers on poles further step down the electricity as it prepares to enter our home**
7. **You can now flick on the light switch and see the energy in action**

So, as you can see, this is a multi-step process which results in energy traveling hundreds and sometimes thousands of miles from its point of origin at the power plant. That inefficiency also results in power loss along the way. Energy's lost as heat while it travels on the transmission lines. When you are traveling, you can get an idea of the amount of heat being lost on these lines by the amount that they droop and sag. The heat causes them to expand over time and so when you see lines hanging especially low, that means they are very hot and very inefficient. Typically, the amount of energy lost from the power plant until your home is about 5%.

. . .

WHAT'S MORE is that if 5% energy loss doesn't sound like much, consider the fact that only about 1/3 of the potential energy from a lump of coal, oil or nuclear even makes it onto the grid as energy. To put that in perspective, the amount of energy we lose because of thermodynamics is more than the annual gasoline consumption in the U.S.

LAST, once the energy makes it to our home, it doesn't stop there. Energy travels to our appliances and if they are not energy efficient, we are likely to see further energy loss, of which we are now responsible for footing the bill. Because once the power hits our home, the power company is no longer responsible for losses, we are. Next chapter, we'll discuss ways to help reduce the power consumed at home.

The True Expense of Energy Production

ENERGY PRODUCTION at scale has efficiency issues that small production at home simply does not. It's no wonder people can reduce their monthly utility burden by switching to renewable energy. Let's look at the costs of energy production at a typical power plant.

FIRST, we have significant generation expenses to account for. These are things like paying workers to extracting the coal or oil, then paying to ship the physical product long distances, and finally a series of workers, managers, and executives are all paid to help keep the energy flowing.

NEXT, we must factor in losses from the wasted energy because of thermodynamics and the production of energy, which can account for almost 70%. Yep, that's almost 70% of the potential energy literally going up in smoke all before it even reaches your home. Then there is the loss of energy from your home to the very appliances you require, the energy to power. In addition, our electric bills also must factor in paying for people to maintain the poles, transformers and lines that deliver the power. These are skilled employees who get paid well and guess what? When a storm comes and the power goes out, they get paid even better to risk their safety to fix it.

LAST, is the environmental costs of traditional energy production. That includes the damage done extracting the energy from the ground as oil or coal. Disposing of the waste which can be extremely dangerous and hazardous, especially if we include nuclear energy waste in the discussion. The pollution in the air during production and sometimes long after. And finally, the pollution that results from long since abandoned factories that no longer have a use and sit vacant as a blight on the land. No matter how you slice it, traditional energy production is an expensive, wasteful process.

What I Wish My Roofer Had Told Me

Why Solar Is a Better Alternative?

MUCH LIKE THE food production was better when done at a smaller scale at a local level, energy production is also better at the small scale too. In the past, every town had several farms delivering food locally and rotating the crops to ensure the land was productive for the next generation. Late last century, they dumped the process of crop rotation in favor of single crop production, leading to higher yields now at the expense of future production and biodiversity. Energy production has grown much the same with large energy production facilities producing huge amounts of power, but suffering from waste and loss as it travels far distances to our homes. In addition, we generate that power at the expense of future generations. It was one thing to do that when we didn't have realistic alternatives, but now we know what the damage is and we also have alternative clean energy sources at similar financial cost, producing a negligible environmental cost.

IF WE CHARGED energy companies based on their true cost, i.e., the damage to the atmosphere, the soil, or even the damage to the enjoyment of a beautiful view, it would make the choice not really much of a choice at all.

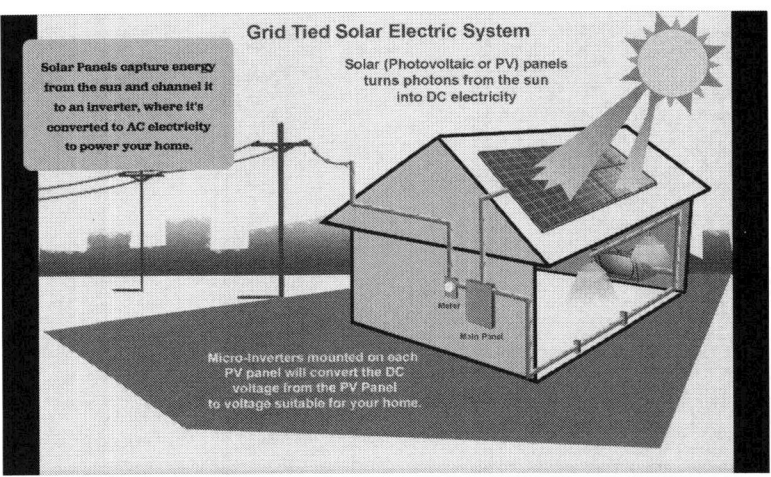

Here are the 18 best reasons that solar makes sense for your finances and your families future

Solar is Cheap

YEP, you heard me right, solar power is a cheap source of electricity to power your home. Assuming you have the right roof and you don't live in the middle of a forest, solar has the potential to offset over 100% of the electricity you use every single year. In fact, you can often receive credits from the electric company for excess energy that you produce.

OVER THE PAST decade the price of solar panels has dropped about 70% as they have become more widely adopted and efficient. Presently, the chief obstacle to a lower price are the materials needed to produce them and as more people switch to solar, the price may actually climb. While a typical system here in the US costs anywhere from $10,000 to $40,000 depending on energy needs and the efficiency of your particular roof, financing has turned that into a cheap monthly expense that can be less than you were paying on your electric bill. What's more is that just like your house, once you've paid off the panels you own them free and clear and any energy they produce is almost FREE.

BESIDES A DROP in the price of solar panels, several states offer tax credits to help further entice homeowners to make the switch. These tax credits will not last forever, but while they are here, we encourage everyone to take advantage of the potential savings they can create. As the US pushes everyone towards green energy and electric cars, early adopters will be rewarded, and those that wait will be required. It's fairly obvious that US citizens are being strongly encouraged to make the switch to cleaner energy. I would rather see you get rewarded for early adoption with tax credits and savings than punished with higher prices as costs rise due to increase demand and new regulatory requirements.

Buy Don't Rent

ONE POINT many people don't realize is that essentially when you switch to solar, you switch from renting your energy to buying it. When getting your energy through traditional energy companies, that monthly payment you make each month goes to pay off their aging infrastructure. With solar, that payment you make each month goes to pay off YOUR infrastructure. You have become your own power plant, and that's an amazing feeling. You are paying either way. Why not get something for your money? To me, this is the most no brainer decision you could make with your money.

More Control Over Your Bill

ANOTHER IMPORTANT REASON TO switch to solar is to stabilize and control your monthly energy costs. If you could have locked in your gas prices at $2.00 a decade ago and know that in another few years, your yearly gas price would be next to free, would you have done it? Switching to solar locks in the rate you pay now, saving you substantial money when the price of electric goes up in the future. You'll no longer be shocked by large monthly energy increases. You can have a predictable monthly investment in your home that won't change. An investment that does not cost more each month and that will not throw unexpected curveballs to your household budget. For me, I hate surprises (just ask my friends the last time they threw me a surprise party) I enjoy setting an expectation and having it met without dealing with the unexpected. Basically, financing a solar project allows me to have a set monthly payment that does not change or surprise me.

Increases the Value of Your Home

IF YOU WERE MOVING to a new house tomorrow, would you want one with solar panels and a low monthly electric bill, or a house with no solar panels and a higher monthly electric bill? Of course, you would want the ability to save money with a solar system and guess what, so would a potential buyer of your home, too. Recent reports state that, on average, a home with a solar system sells for over 4% over a comparable home without it. What's more is that a solar powered home also sells faster, which in a competitive market can make a big difference. So not only do solar panels add value to your home and save monthly electric costs for you and the buyer, but they will also help your home sell faster!

Potential Battery Backup

IF YOU LIVE SOMEWHERE that suffers from frequent power losses, a battery backup system that is connected to your solar system might make a big difference to your quality of life. All over the United States, areas suffer from both aging infrastructure of the power grid and frequent natural disasters that can knock out electricity. Losing electricity can cause your food to spoil, make it impossible to flush toilets, cause your home to be so cold/hot it becomes uninhabitable, or just make your living environment less safe. That's not to mention that in some areas of the US criminals take advantage of power losses that knockout cameras and security systems and look to break into homes. Nothing feels better than having power when others do not and being able to help friends and neighbors in need.

The Solar Savings Start on Day One

Unless you pay with cash (and good for you if you do), switching to solar can start saving you money from day one. Once installed, solar panels immediately begin generating power from the sun and helping to offset and often cover your electric bill. The worlds filled with investments that take time to produce results. Lucky for us, solar is one of the few that start working for you immediately.

Passive Income Baby

Whether you want to call it income or savings, you were paying a bill before and now put that same money towards a home improvement. Heck, sometimes you put that money towards an investment in your home AND still saving on your monthly energy bill. Saving money is always better than earning money because:

- **You don't pay taxes when you save money like you do when you earn additional money**
- **Savings basically do not change and are consistent**
- **By making the sun do the work, you enjoy a passive income that can offset energy bills and pay off the solar panels**
- **Most times, you can use the savings to help pay down debts such as car payments or house payments, meaning that you might save years off your mortgage just by using money you were previously paying the utility company.**

This is one area I am most excited about! As a graduate of finance, I love seeing ways that people can protect their bottom line with no risk to themselves. Investing in the stock market has risks. Speculating in real estate carries risks. Reducing your monthly utility bill, well, that's about as risk free as an investment comes!

Solar Power Comes With a Warranty

When you agree to have solar panels placed on your home, you would expect someone to stand behind the product you are investing in, right? Well, the good news is that most solar companies warranty their product anywhere from 10 to 30 years. So assuming you find the right company, you can expect to get a product and still have parts and labor covered long after it has already paid for itself. That's a pretty fantastic deal if you ask me.

Have a warranty question? Visit jonnelsen.com to get connected with a 30 year warranty with one of the largest and fastest growing residential solar companies in the nation.

Decrease Dependance on Foreign Oil

Depending on the current administration in the White House and the party in charge of congress, the US imports up to 50% of its oil from foreign sources. While that fact might not surprise many people, it might surprise them to know that the United States is often the top oil producer in the world. So increasing the national investment into solar energy has the potential to help eliminate the oil we import from foreign sources and allows us to invest in local generation of oil and natural gas.

Solar Is Obviously Great for the Planet

EVERYONE KNOWS solar is great for the planet and our future, but do you know why? The biggest reason is because in the US much of our power comes from coal. Coal power plants are the single most expensive and dirtiest forms of energy production exist. It takes over thousands of pounds of coal to power the average US home for a year! Residential rooftop solar production can eliminate the toxic coal particles from entering the atmosphere and our lungs! This is one reason I am passionate about solar energy. I want my son to grow up in a world that is cleaner than the one I grew up in, and I hope he has the same goal for his children as well.

Traditional Energy Travels a Great Distance and Suffers From Energy Loss Along the Way

AS WE PREVIOUSLY DISCUSSED, the United States has three energy grids that power the country. They are the Eastern, Western, and Texas Grids. With only three potential grids to draw power from, that means that the energy travels a significant distance across the land to get to your home. On average, about 5% of the power that is produced is lost simply by traveling through the wires before it reaches your home. That is not to even account for the cost and spent energy needed to power the trucks, buildings and storage facilities associated with traditional energy production. Now with solar, the power produced travels instantly into your house, resulting in almost no loss and when your rooftop generates excess energy, it's sent to the grid. That excess is measured at your house before it enters the power lines, meaning you get full credit for what you produce.

Solar Panels Are Extremely Durable and Almost Maintenance Free

A COMMON MISCONCEPTION is that solar panels are very fragile. While that is partially correct because the actual solar panel is very fragile, the glass and frame that enclose those panels are extremely durable and built to withstand punishment from the elements. Most panels can handle any extreme wind, hail, and debris that Mother Nature can throw at them. In addition, they are easy to maintain and might only require a basic cleaning once or twice a year to help keep some of the dust and dirt off the glass. While that basic cleaning is truly unnecessary to their function, it will lower efficiency over time if not removed. Most homeowners will find that rainwater does most of the work in keeping the panels relatively free of dust and a homeowner or solar professional can do a simple annual cleaning easy enough. Most times, I recommend you call the professionals as they're insured against any issues and have the experience and tools to do the job quickly and correctly. If you are determined to save the money and do it yourself, there are many YouTube videos that help walk you through the process, and we will discuss it later in this book as well.

Solar Energy Is a Job Creator

CURRENTLY (2022) THE United States solar industry is directly responsible for about 250,000 local jobs and indirectly responsible for countless more. While many industries are downsizing or laying off workers, solar is expanding as more homeowners and companies seek to take control of their future. If you wish to join the solar revolution and be part of a growing industry, you can visit jonnelsen.com and I will be happy to get you connected with a better future in solar. Opportunities are endless, whether you are looking for a side hustle, part-time, or full-time employment. And if you are a homeowner making the switch to solar, you can feel good about the fact that the decision you made to go solar directly affected several people in your neighborhood and across the US.

Solar Takes Advantage of Underutilized Space

THAT ROOFTOP of yours doesn't really have anything better to do, right? Or perhaps you want to keep your roof free, and would rather focus on underutilized land on your property? Either way, you can benefit from the flexibility of solar to generate revenue on your property using land and rooftop that would otherwise go to waste.

Solar Panels Keep Your Rooftop Cooler

ONE FREQUENTLY IGNORED benefit of solar panels is the fact that they help reduce the amount of direct sunlight reaching your rooftop and therefore reduce the overall temperature of the roof. This can translate to significant energy savings during those hot summer months. Capturing that sunlight turns the sun from an enemy to an ally as you seek to reduce energy bills. Put another way, simply having solar panels on the roof collecting that scorching summer sun might allow you to reduce the time you have the AC blowing and save additional money each month. This is especially useful to those who live in warm climates and use the AC often.

Solar Panels Can Work Even Better in Cold Weather

BELIEVE IT OR NOT, cold weather is actually a great time for solar panels. Short of being buried under snow (which will melt faster with panels underneath) that winter cold helps the panels function even more efficiently. When there is snow on the ground but the panels are clear, homeowners benefit from The Albedo Effect, so the light color of the snow helps reflect sunlight up rather than being absorbed into dark surfaces.

Solar for Off Grid Use

ONE POPULAR FUNCTION of solar is the ability to generate your own power and thus be able to remove yourself from the grid. While I don't recommend this, as there are many benefits of being associated with the grid and its infrastructure, it's the hope for some that they can remove themselves from reliance on society and be more self-sufficient. Solar panels and a strong battery backup system will be helpful in creating energy during peak hours and distributing it when you need it after dark.

Using the Existing Utility Infrastructure for a Small Monthly Fee

WHILE A SELECT FEW enjoy the idea of being "off the grid", for most people, connectivity is important. Being part of the grid offers many benefits and sometimes allows you the ability to receive energy credits when you generate excess power. To stay on the grid, it can cost as low as $10 a month and grants you access to everything you had while paying a larger energy bill. This protects you from ever being without power, even while waiting for a repair or replacement should something ever happen to your panels.

THERE YOU HAVE IT FOLKS, a good number of reasons it makes sense to switch to solar. While this is not an exhaustive list, as I know there are plenty of more reasons someone might make the change, this covers most of the largest reasons. But you know what, every situation is different and maybe you have your own reason for wanting to take control of your energy. That's ok, when speaking to a solar consultant they can help you understand if it makes sense for your family to make the switch and if solar will truly solve the problems you want it too.

panels + system

...

How Solar Panels & the System Are Designed and Function

"Once you got a solar panel on a roof, energy is free. Once we convert our entire electricity grid to green and renewable energy, cost of living goes down."
  - Elizabeth May

WHILE SOLAR PANELS SEEM COMPLICATED, the actual way that they convert solar energy to electrical energy that we can use is anything but. Solar panels are actually made up of many smaller units called photovoltaic (PV) cells or, as most people refer to them, solar cells. PV cells are composed of a semiconductor material, which is typically silicon based. Now semiconductor simply means that it can conduct electricity better than an insulator can, yet not as well as a stronger conductor like metal can. Silicon is one of the most abundant elements on earth.

How Do Solar Panels Convert Solar Energy to Electrical Energy

Now, in a solar cell, crystalline silicon is layered between two highly conductive layers. Silicon atoms are connected to these neighboring atoms with strong bonds. Now within the silicon cells are different silicon types, each with specific properties and abilities. Because of those differences, they can create both positive and negative charges. The photons from the sun's rays hit the silicon and travel on the top layers of the cell, forming an external circuit. Each specific silicon cell on the panel creates about a half volt of electricity and in a panel of 12 (PV) cells that could create enough energy to charge a typical cellphone. Now multiply these panels many times over and you could create enough electricity to power an entire home.

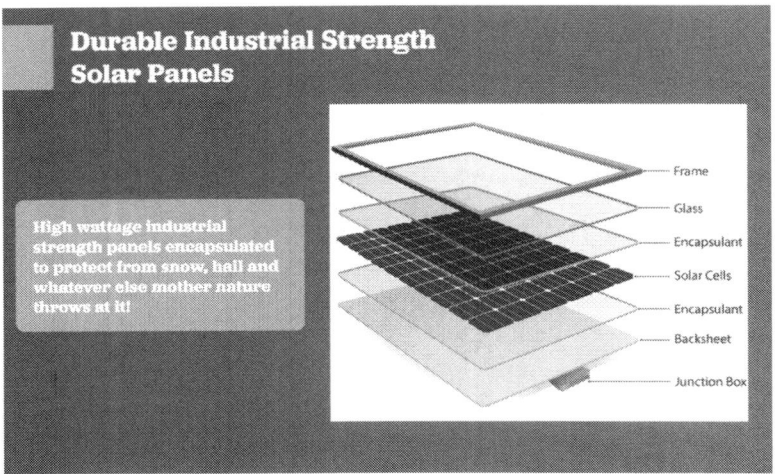

Durable Industrial Strength Solar Panels

High wattage industrial strength panels encapsulated to protect from snow, hail and whatever else mother nature throws at it!

- Frame
- Glass
- Encapsulant
- Solar Cells
- Encapsulant
- Backsheet
- Junction Box

How Does Residential Grid Tied System Electric System Work?

WITH A GRID TIED SYSTEM, there are often many misconceptions. Among them is the idea that a battery must be installed with the system and, in most cases, that's unnecessary. In most modern systems, a battery is unnecessary, because the utility company acts as a giant battery. That means that during peak hours, your system will purposefully create more power than is necessary and send it off to the grid through the power lines connected to your home. During the evening, when the solar system isn't generating energy, you will draw from that same grid and take back the excess energy your home created.

A FEW PIECES of equipment that help make the entire system work are the inverter and meter. Both connect to your home and the grid. The sun will hit the solar panels and the inverter will then change the current from D/C to A/C power. A/C current powers your home to power and the various electronics and appliances within it. The principal goal of any solar system on your roof is to generate not only what you need for your current usage today, but also what you might need in the future. With the right solar installation, you no longer need to keep the house at uncomfortable temperatures during summer and winter months just to reduce your energy bill. In fact, even if you don't have a need for all the energy you produce, most utilities will offer you a credit for an excess production that makes it to the grid from your home. No one invests in a home solar project to scrimp on the energy they use. Homeowners want to enjoy the benefits their home offers and plan for any additional future use.

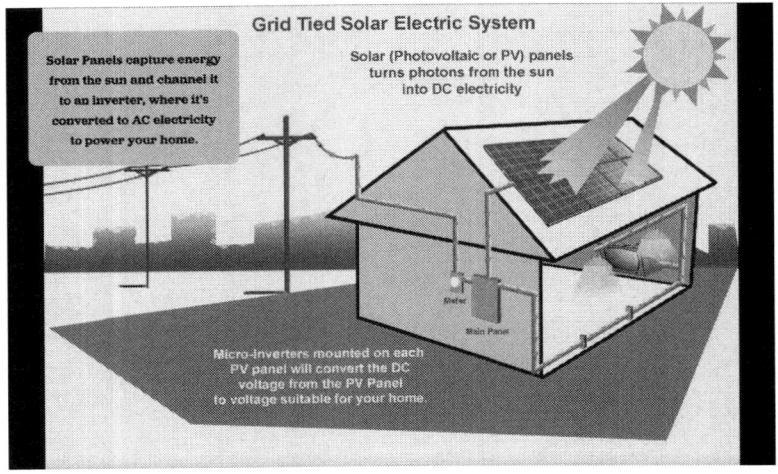

Could the Entire Planet Run on Solar Power?

THE SUN IS the most powerful force in our entire solar system and creates enough energy to power everything we need multiple times over. Hypothetically then we should be able to power all our electricity needs from the sun, right? Well, not so fast. Besides any political factors that might stifle complete renewable solar energy, there are additional factors. One of these factors is the strong economic benefits that many businesses and countries like Russia have from maintaining a dependance on fossil fuels like coal and oil. For instance, even here in the United States there's a powerful well funded oil lobby that has significant sway in congressional decisions. This lobbying group spends millions each year helping to elect politicians sympathetic to their long term financial interest.

And this is not a knock on politicians, because who really likes them anyway? Many of these politicians are well intentioned enough and help prop up these aging business models because people that they represent work in these industries and any transition away from the money they generate would create a hardship for their constituents. No politician wants to take good paying jobs away from the people who vote for them. That is not a recipe for a long life in politics.

BECAUSE OF THE POLITICAL FACTORS, you will often see significant renewable energy support from senators in states that do not have a large base of constituents dependent upon the fossil fuel industry. Conversely, the politicians from states like Texas, Alaska or West Virginia, for example, have a vested interest in protecting the citizens of their state and the jobs that puts food on their families' table. So no matter how dire a situation some may feel about climate change, it's not always fair to paint with a broad brush, especially when you consider how you might feel if your friends or family members suffered immense financial hardship because of changing energy policies. One significant benefit of solar is that it is a fast-growing segment and is responsible for millions of jobs across the US from research and sales all the way to construction workers and installers.

ANOTHER REASON it is not reasonable to expect worldwide solar adoption soon is because of the physical and logistical barriers to a fully solar powered planet. One obvious reason is that sunlight is unevenly distributed around the globe. Places like Norway and Alaska have towns in their northern regions that can go weeks or months with no sunlight at all. Without another way to generate electricity, those areas would become uninhabitable during winter. To power those areas with the sun, electricity would need to be sent hundreds or thousands of miles from sunny areas to ones that are not receiving adequate sunlight to power their needs.

CLOUDS and overcast weather certainly reduce the effectiveness of solar panels to generate the electricity they're capable of producing. That is why homes that choose to go off the grid completely have large battery storage units to save that previously generated solar for a rainy day. Houses that stay on the grid (which I recommend) have the ability of utilizing the electric companies' existing grid as a giant battery, producing excess electricity when it's sunny and drawing on it when needed during nighttime and inclement weather.

THE LAST MAJOR limitation is one of technology. Presently, even the most advanced solar panels researchers have constructed convert less than 50% of sunlight into usable electricity. These research grade panels exist mostly as experiments and concepts, rather than commercially available systems. The commercial systems available to homeowners are about half as efficient as those, converting a maximum of 25% of sunlight to energy.

DESPITE ALL THESE LIMITATIONS, if funding, land and resources were available tomorrow, current technology could indeed power the entire world. It would take hundreds of thousands of square miles in an area like a desert. To put that in perspective, the Sahara Desert itself is over 3 million square miles. Recently, countries like China have invested millions in research and implementation of massive floating solar farms in the ocean that are creating enough energy to power cities.

. . .

THERE ARE hundreds of millions of people living around the globe that currently do not have access to stable energy sources and, for those, solar is the easiest, cheapest, and best solution for reliable electricity. In the US, despite the inconveniences we often face because of the aging energy grid, we're blessed with a reliable source of abundant energy. And every house that makes the choice to transition to solar helps make that grid stronger by taking pressure off it during times of peak usage.

EVERY DAY, more and more people are making the switch to solar and are effectively creating mini power plants to power not only themselves but their community, all because of the ability to stay connected to the grid and lend power to the utility companies. The best part is that the extra energy your system creates returns to you as a credit.

is my house right for solar?

. . .

Find Out if Your House Makes the Cut

"Have nothing in your house that you do not know to be useful, or believe to be beautiful."
- William Morris

AS A SOLAR CONSULTANT, it might sound surprising for me to say, but not everyone should make a jump into a residential solar system.

Not All Homes Qualify for Solar Energy, But Your's Might

SOMETIMES A RESIDENTIAL SOLAR system just makes little sense either financially, logistically or both. This chapter will discuss some reasons solar might not be the best investment and help you decide if you are one of the unfortunate few.

What I Wish My Roofer Had Told Me

Does Your Area Receive Enough Sunlight?

THE FIRST AND most important reason solar panels might not work at your home is based on the amount of sunlight you receive during the day. One common misconception is that solar panels need heat, which is wrong. In fact, heat can often make the panels less efficient. Temperature is not a big concern one way or the other, however the amount of direct sunlight is. All panels need sunlight to function optimally, but when looking at a roof's efficiency, it's best to consider an entire year versus day by day or month by month efficiency. For instance, during the winter months up north, the days are traditionally shorter with less sun, but in the summer sun more than make up for that.

GOOGLE HAS MADE the process easier for certain homeowners across North America with their innovative "Project Sunroof".

Google uses satellite imaging to predict factors like hours of usable sunlight per year and square footage on your roof that would work for panels. In fact, they even tell you how many trees would need to be planted to equal the savings your solar roof is generating annually. While much of the data is based on estimates, it is helpful to review the information if you live in a region that "Project Sunroof" has reached.

CLOUD COVER ALSO PLAYS a significant role in the solar efficiency of your roof. While a solar panel still works and produces electricity on a cloudy day, the total amount is diminished. Cloudy days can reduce production anywhere from 50%-90%. When your systems designed, your solar rep will help you understand the pros and cons of the area and what to expect seasonally. They will then work to ensure that your system performs as it needs to for your needs.

OFTEN PEOPLE HAVE some concern about rain or fog interfering with the stability of the panels, but that won't be a problem. Solar panels are well-sealed and will not have any mechanical issues from rain. In fact, rain is helpful in keeping dirt and dust off the panels, so they perform their best. Rain only ever becomes an issue in the rare event a panel has cracked or broken. If you notice a broken panel, it's best to notify your solar installer as soon as possible to get it repaired or replaced.

ANOTHER BIG ISSUE with solar panel efficiency is snow. If you live in a place that receives for months on end like Buffalo or Minneapolis don't worry while you won't have the same energy generating ability as someone living in San Antonio, there's still an excellent case for you to make the solar switch. For one, while snow cover prevents the panels from generating power, it's not impossible to remove the snow and clear the panels without the help of a professional. Here is a list of tools that homeowners can use to remove snow from the panels:

- **Roof Rake** - These are specially designed to remove snow, ice and debris from a home's roof. In fact, even if you don't own solar panels, these can help take some weight off the roof and allow the snow to melt faster. Use common sense with the proper roof rake. If you wouldn't use it on your car, it's best not to use it on the panels.
- **Soft-Bristle Outdoor Broom** - While perhaps not as effective as a roof rake, soft-bristled outdoor brooms will be safer on the glass face of the panels. Be sure to look for one with a telescoping handle to allow you to have a distance between you and the falling snow and ice.
- **Leaf Blower** - A normal, everyday leaf blower does a great job of removing fresh and fluffy snow from the panels. This method will most likely require you to use a ladder and for that you will want someone helping to steady it, especially with the force of wind pushing you from the blower. Another point of caution should be to ensure any electrical cords stay away from any wet snow or puddles.
- **Spray with a Hose** - If the weather is warm enough, it's worth trying to spray with lukewarm water from a hose. A few minutes should be enough to clear the snow from the roof and get you back to generating big time power in no time.

- **Tennis Balls** - Sometimes all that's needed to send the snow floating down the panels and into the yard is a well-placed throw of a tennis ball. If you have decent aim, a few gentle tosses might be enough to clear your panels. This trick won't work as well with heavier snow and you may end up getting the balls stuck up there instead.
- **Heating System** - Heating systems are specially designed for solar panels and are very common in areas that receive higher than average snowfall each year. They typically contain a series of small pipes or hoses that are filled with warm water. Sometimes they're manufactured with electrical wires or coils that attach to the panels themselves. Either way, while costing more than a typical garden hose, these will make the job of rooftop snow removal that much easier.

AT THE END of the day, snow is not a make or break for potential solar production, it just might mean taking proactive steps to ensure you remove the snow quickly to keep your system generating. Each week, more and more homes across the north are making the move to solar and that's a trend that will continue. Heating and cooling a home using power generated with solar is a great option if paired with a failsafe system like a fireplace or electric heaters that can warm a larger space in a pinch. By utilizing battery backup systems, you can ensure that even during a storm, you have power to spare if your goal is to avoid drawing from the grid. If you live off the grid, a fireplace for warmth is a necessity in times of prolonged snow storm and snow cover.

Does Your Particular Roof Receive Enough Sunlight?

WITH SUNLIGHT, there are more factors to consider than just the natural sunlight of your region. Factors like tree cover in your yard or a neighbor's yard can affect the sunlight that reaches your roof. Chimneys, vents or other obstacles can also cast shade and make the solar panels less effective. If you go to the trouble and expensive of adding solar, it might be worth considering if removing a tree or two to generate more annual production is the right move. Perhaps only portions of your roof wouldn't work for solar, but you could offset a portion of your energy use with a smaller system. With so many different options, there are plenty of ways to make solar work.

THE ORIENTATION of the roof is also a factor for solar effectiveness. In North America, houses with southern exposure will collect the most sunlight, however eastern or western exposure can also produce results. If you have northern exposure, your ability to collect sunlight is reduced and you will have to consider if the benefits outweigh the cost. Ask your solar rep to review your homes orientation to establish if and where solar panels would be most effective.

IN THE SOLAR WORLD, size matters. The size of your roof will determine the amount of panels it can support and if you have a large enough area to completely offset your annual electricity needs. A smart rule of thumb is to fill up all usable space with solar panels and if you generate additional electricity, you can often receive credits from the utility company for them. Most solar installers suggest a design accounting for 115% of your current electricity usage and if you plan to get an electric vehicle in the future, you may want even more. The last thing you want is to install an expensive solar panel system and still have to worry about running the AC on a hot day. Always overestimate what you will need in the future so you can feel comfortable no matter the season or temperature. That goes double if your home has central air and you had kept the temperature in the house uncomfortable to save money in the past.

What Condition Is Your Roof In?

BELIEVE IT OR NOT, the type of roof you have also plays a key factor in determining the potential efficiency of residential solar. Depending on where you live, certain roof types may be more popular than others. Here in the North East, most people have asphalt shingle roofs which work great for solar installation. Metal roofing also is amazing for solar panels because they already have seams to connect and drill into. In regions like the South and South-West, tile roofing is more common and that can be trickier. This is because of natural tile brittleness and the potential for damage to the water-proofing barrier beneath them. However, solar installers are highly motivated to make it work and as they say, "where there's a will, there's a way". As long as the contractors say it can be done and will back up that claim with a written warranty, it shouldn't be an issue no matter what type of roof you have. The exception to that, is a wooden roof because of the possibility of serious fire hazard. If you have a wooden roof, but still want solar, your best bet would be to consider a ground-mounted system in your yard.

. . .

Do you know the specific angle of your roof? Chances are you don't, but with solar 30-degrees is the optimal tilt. However, no matter if you have a flat roof or live in an A-frame solar can still take advantage of a solar roofing system if your installer feels comfortable adding brackets to tilt them (which is a common solution). You can also sacrifice a bit of efficiency and mount them flush if aesthetics are your largest concern. As long as you can work the angles and get a reasonable tilt towards the sun, your system can begin generating energy on day one.

How New Is Your Roof?

UNLESS YOU HAVE RECENTLY PUT a roof on your house, chances are high that your solar panels may outlive the roof they're placed upon. At present, solar panels can last up to 40 years and we may find out that they can last even longer because of advances in technology, improved components and fierce competition forcing increased quality. While a quality roofing company will usually offer a 30 year product warranty, a quality solar company will usually offer the same. Is your roof fairly new? Then now is definitely the time to investigate if solar makes sense for your family. If you need a new roof, reach out and see if you can have both installed at the same time. This might sound like a headache, but has the potential to save you a bit of money and time in the long run.

DEPENDING on your current electric bill, most residential solar systems can be fully financed. Once financed, return on investment can happen in as little as 5-10 years, meaning that after you finished paying for it, you own the system outright. My rule of thumb is if you expect your roof to last as long as it takes to finish paying off the system, you're good to go. Roofing is a competitive business and companies are often willing to negotiate with you to get the job done. If you need a new roof, your first call should be to the company that installed the system in the first place to see if they warranty removal and replacement of a system. Most will not, but it's always worth asking. Another option would be to negotiate with the roofing company for removal and reinstallation of the system upon completion of the job, assuming that they have experience with that type of work. Otherwise, you might need to pay a small amount out of pocket to have it removed and reinstalled, but that is a drop in the bucket compared to your savings over the years from a reduced energy bill. That is especially true once the system is paid off and generating power without a monthly payment.

Are There Any Incentives Available?

INCENTIVES whether at the national level, regional level, or state level can also help reduce the cost of solar. There is no guarantee that there are available incentives, but everyone should always do research into any potential savings. The following websites can provide valuable resources to potential tax savings:

- US DEPARTMENT OF ENERGY (ENERGY.GOV): Here you can read articles about ways to reduce your electric bill and information about potential renewable energy sources. While this book focuses on solar production, there are many other ways to produce power that the average homeowner might consider. At the top of the page, there is a simple search bar that can provide information about things like "Clean Energy Tax Credits".

- N.C. Clean Energy Technology Center at N.C. State University (dsireusa.org): DSIRE allows you to search by zip code and discover any potential incentives available in your region or state. This is the most comprehensive list available, but you will need to carefully read each incentive for potential viability. Some incentives are for corporations, special groups or certain states. There are tax credits, loan programs, grant programs and rebates, all listed and organized. If you apply a filter to the search criteria, you can ensure that what you see meets your unique situation.

If you want all the links consolidated in one place, visit jonnelsen.com. However you decide to access the incentives list, just make sure you do so you don't miss out on valuable opportunities that could save you **THOUSANDS.**

also by jon nelsen

- *Starting a Bed and Breakfast: Bite Sized Interviews With Successful B&B's on Building a Brand That Lasts*

- *Running a Bed and Breakfast: Bite Sized Interviews With Successful B&B's on Maintaining a Thriving Inn*

- *One More Beer, Please (Vol. 1, 2, 3): The Largest Collection of Interviews With Brewmasters and Craft Breweries*

- *Solar Panels: Are Solar Panels Worth It?*

- *Complete Guide to Roofing and Solar: Homeowners Essential Handbook for Money Saving DIY Roof Construction and Solar Panels*

- *What I Wish My Roofer Had Told Me: The Ultimate Guide to the Roof of Your Dreams on a Budget*

- *The Only Time I Set the Bar Low Is for Limbo: Reaching Your Potential in Work, Life, and Relationships*

- *What College Didn't Teach You About Getting Hired: The Ultimate Guide on How to Find a Job After Graduation*

- *The Anxiety Answer: The Step-by-Step Guide to Overcoming Fears, Phobias, and Other Voices in Your Head*

- *Solar Powered Energy Theft: Legal No Money Down Solar Panels for Homeowners*

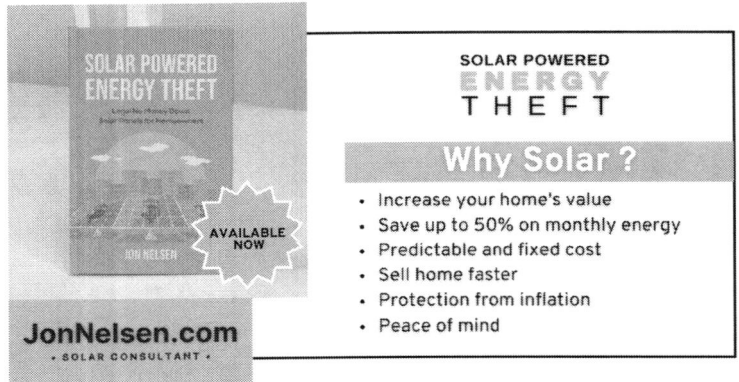

AVAILABLE FREE FOR A LIMITED TIME

Printed in Great Britain
by Amazon